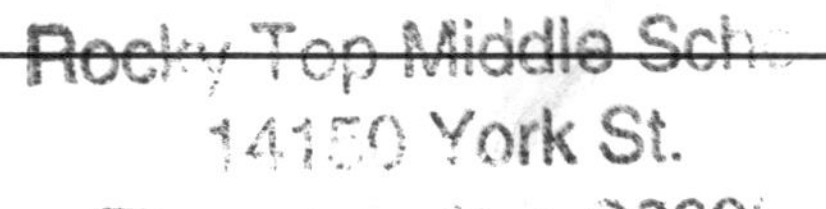

Spiraling Through Life with FAST PLANTS

An inquiry rich manual

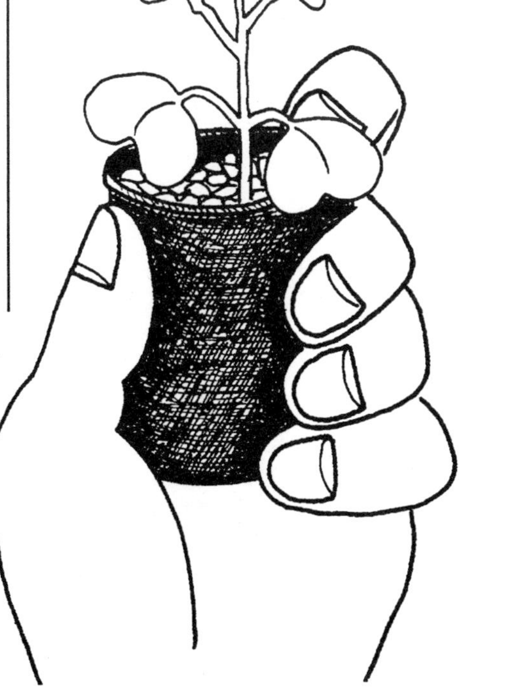

KENDALL/HUNT PUBLISHING COMPANY
4050 Westmark Drive Dubuque, Iowa 52002

College of Agricultural and Life Sciences, 1630 Linden Drive, Madison, Wisconsin, 53706
www.fastplants.org • wfp@fastplants.org • 1-800-462-7417

Teachers may copy selected pages for student use.

To order copies contact:

Kendall/Hunt Publishing Company
1-800-228-0810 (phone) • 1-800-772-9165 (fax) • www.kendallhunt.com

Revised Printing 2004

ISBN 0-7575-1137-6

Printed in the United States of America
10 9 8 7 6 5 4 3 2

Authors: Robin Greenler
John Greenler
Daniel Lauffer
Paul Williams

Illustrator: Amy Kelley

Wisconsin Fast Plants Program: Paul Williams
Coe Williams
Daniel Lauffer
John Greenler
Robin Greenler

We would like to thank Wayne Becker, Steve Smriga, Chrissy Rikkers, Jim Lucey, Whitney Hagins, Rick Berken, Judith Fischer, and Kandis Elliot for their assistance, editing, and support.

Fast Plants Materials and Supplies

This manual is designed to employ locally available low-cost materials. The Fast Plants seed and alternative growing systems with support materials are available through Carolina Biological Supply Company, 2700 York Road, Burlington, North Carolina, 27215
www.carolina.com • carolina@carolina.com • 1-800-334-5551

CONTENTS

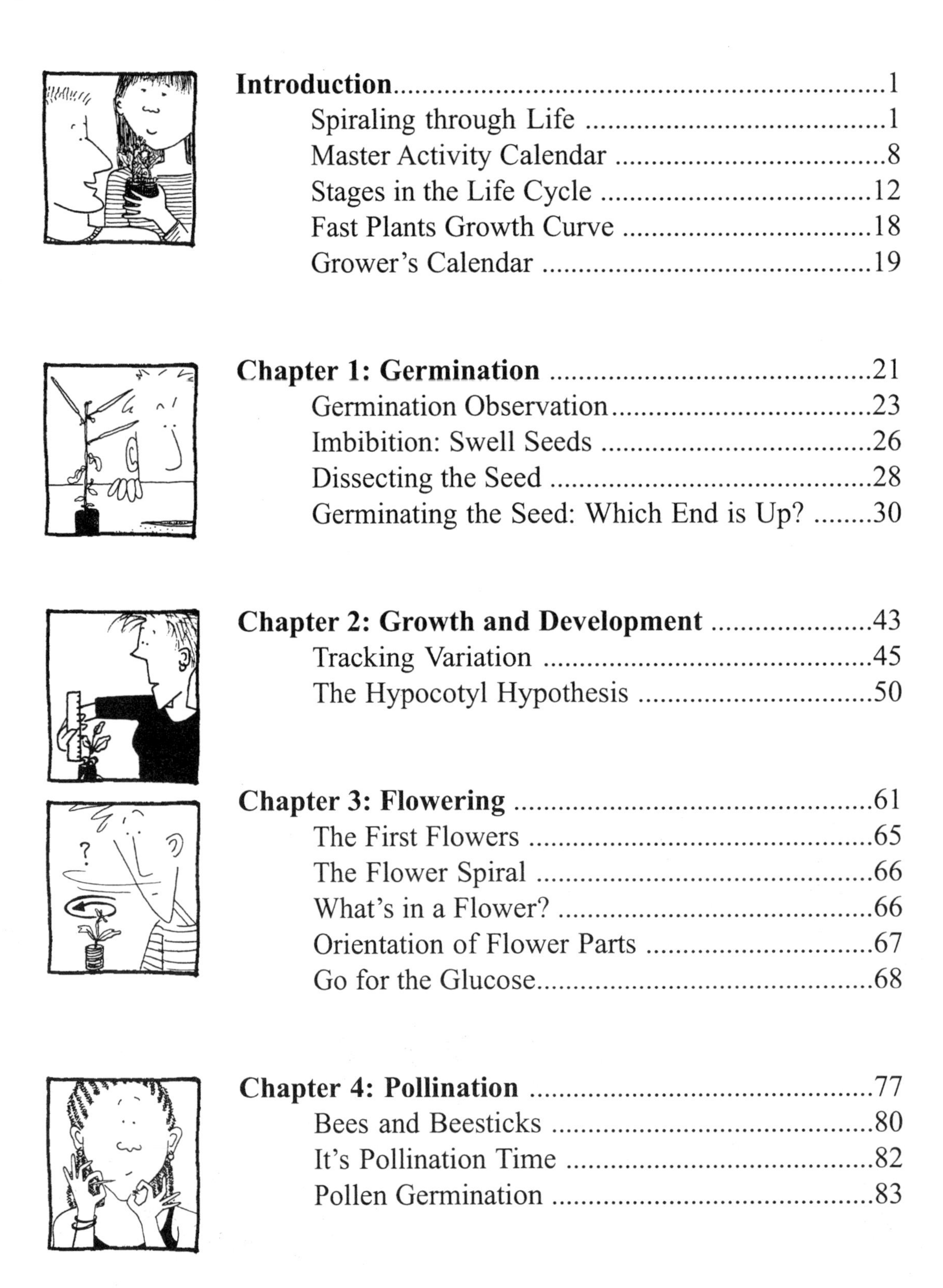

Introduction............1
Spiraling through Life 1
Master Activity Calendar 8
Stages in the Life Cycle 12
Fast Plants Growth Curve 18
Grower's Calendar 19

Chapter 1: Germination 21
Germination Observation............23
Imbibition: Swell Seeds 26
Dissecting the Seed 28
Germinating the Seed: Which End is Up? 30

Chapter 2: Growth and Development 43
Tracking Variation 45
The Hypocotyl Hypothesis 50

Chapter 3: Flowering 61
The First Flowers 65
The Flower Spiral 66
What's in a Flower? 66
Orientation of Flower Parts 67
Go for the Glucose............68

Chapter 4: Pollination 77
Bees and Beesticks 80
It's Pollination Time 82
Pollen Germination 83

Chapter 5: Fertilization to Seeds 91
Embryogenesis—What, When and Where? 96

Epilogue 109

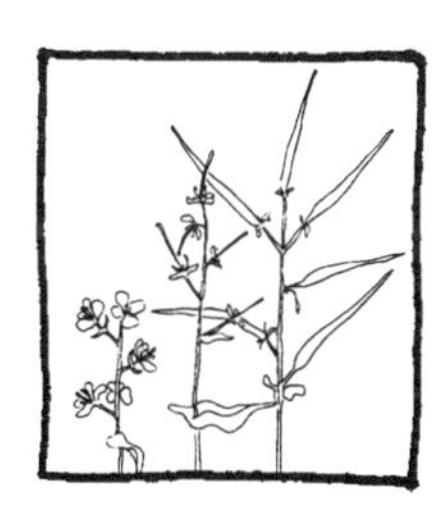

Appendices 113
Appendix 1 Growing and Maintaining 113
Appendix 2 Variation: Collecting, Organizing, and Interpreting Data 127
Appendix 3 Film Can Hand Lens 133
Appendix 4 Scale Strips, Drawing to Scale, and Calculating Magnification 135
Appendix 5 Classroom Vignettes 139
Mr. L., Ninth Grade 140
Ms. H., Ninth Grade 143
Mr. B., Seventh Grade 145
Appendix 6 Science Standards 151

General References 153

Glossary 155

Index 161

INTRODUCTION

Spiraling through life with Fast Plants

A living organism starts as a single cell and becomes a mature, complex plant or animal that is a springboard for a whole new generation. How does this amazing transformation occur? Through exploration with Fast Plants you can gain a deeper understanding of these questions and engage in inquiry-rich scientific exploration.

Fast Plants are a variety of *Brassica rapa.* Familiar vegetables such as broccoli, cabbage, turnips and cauliflower are also brassicas. Fast Plants were initially bred at the University of Wisconsin–Madison as a model organism to study the genetics of this important group of plants. With a life cycle of 35 to 45 days, small size, and simple growing needs, Fast Plants quickly became a popular teaching tool that is used internationally in pre-K through college classrooms.

Spiraling through Life with Fast Plants provides a wealth of background information, activities, and ideas. Through growing a Fast Plant and understanding its

Why Fast Plants?

Fast Plants may be small, but they provide a host of dynamic classroom opportunities. They have been used to guide kindergartners, high schoolers, and undergraduates alike into a better understanding of the scientific process of exploration and experimentation. They also make superb models for studying biological principles on the molecular, cellular, organismal, population, and ecosystem levels.

- With their small size, simple growing needs, and rapid life cycle, Fast Plants are very *classroom friendly.*

- Fast Plants foster a strong sense of *student ownership.* Because they are small, easy to observe, and easy to handle, each experimenter can have her or his plant growing in their own film can pot. Students plant a seed, nurture a plant and it becomes their own.

- In its 35–45 day life cycle, the Fast Plant rapidly passes through the life stages of *growth* and *development.* From germination to embryogenesis and senescence, Fast Plants are a model for the study of the cycle of life.

- Each individual Fast Plant is observably different from others in a population. This variation, which is seen in numerous traits, provides a pathway for introducing *quantitative biology, genetics, biological diversity,* and *evolution.*

- Due to their rapid growth, Fast Plants respond quickly to environmental changes. This makes them an excellent organism for the study of *ecology.*

- With their ease of growth and manipulation, Fast Plants work well for the study of *anatomy* and *physiology,* or form and function. For example, tropic responses to stimuli such as light and gravity can be easily observed within hours.

life cycle, the foundation is set for students to launch into experimentation and inquiry-based learning. Opportunities for experimentation are provided in the manual's "inquiry strand." Running in parallel with the background and activity text, this inquiry strand includes questions sparked by the text, relevant contemporary research that relates to the discussion, quotes, food for thought or ideas for further experimentation. Hopefully, these pieces will provide inspiration, ideas, and approaches to students who want to take an idea further, experiment, and "think outside the box."

Each chapter in the manual focuses on a single stage of the Fast Plant life cycle. Students can take a plant through it's entire life cycle following all activities or focus on a single chapter or even on a single activity. While each activity stands on its own and can be performed in isolation, the completion of more activities in many or all chapters will lead to a deeper understanding of how the various portions of the life cycle interact and interrelate. Further, the chapters focus on skills including observation, experimentation, and data analysis.

Classroom Vignette: Mr. L, 9th Grade

The first science class after Christmas is unusually full of energy.

Mr. L.'s students are informally surveying their classmates to see which research project they are interested in doing. Mr. L.'s second semester classes are not unlike other introductory biology classes where the topics of genetics, protein synthesis, evolution, taxonomy, and the like assume major curricular importance.

However, Mr. L. wants to go further.

He wants to design an educational setting, in the confines of a freshman introductory biology curriculum, where his students can demonstrate their ability to apply the information and experiences they gained in the first semester to a new educational situation in the second semester. The students will build on their first semester experience of growing Fast Plants to model the actions and thought processes of a scientific researcher.

The chatter in the room today tells Mr. L. that these additional challenges of designing and conducting research utilizing the Fast Plants have more than a few students a little nervous....

See Appendix 5 (page 141) for the full vignette.

The chapters are as follows:

Germination starts with the structure of a seed and goes through the first days of the life of a new plant: the seed swells, a root and shoot emerge, and the cotyledons become photosynthetically active.

Growth and Development follows Fast Plants from the seedling stage through to flowering. As each plant increases in size and complexity, there is a concurrent parallel increase in variation from one plant to another. These changes provide the opportunity to explore the development of an individual plant and to analyze diversity within a population.

Flowering covers floral anatomy and the development of flowers on Fast Plants. Flowers provide a key opportunity to hone observational skills.

Pollination opens up exploration into the connection between plants and insects. Their interactions facilitate the transfer of genetic information from one plant to another. This chapter also sets the stage for discovery of the adaptations of both the brassica and the bee that demonstrate their symbiotic relationship.

Fertilization to Seeds brings the Fast Plant back to the beginning of the life cycle. After pollination, fertilization yields a new and genetically unique single cell. Embryogenesis takes this single cell and sees it through to a seed.

> *A powerful way to use a model organism such as Fast Plants is to pursue it in depth over time, and over a spiralling series of ever-enriching encounters. In this fashion, school teachers will be addressing one of the central challenges in education, namely that of providing sufficient depth to the students' life long learning spiral. Out of this a true understanding will result.*
>
> *A spiral of learning through a spiral of life.*
>
> —Paul H. Williams

The Epilogue brings it all around again. Working within the context of a "family" of plants, students can explore the genetic and environmental influences on variation and diversity.

In addition to the chapters, you will find information to help you work with Fast Plants and utilize scientific techniques for experimentation and data analysis. The Introduction presents the basics of Fast Plants. It also discusses how to use this manual in the classroom and provides schedule organizers to help develop a teaching plan.

The appendices contain a wealth of additional information. There are detailed instructions on how to easily grow and care for Fast Plants, how to make your own tools, and statistical instruction related to the activities in this manual. A set of three vignettes provides examples for successful classroom use of Fast Plants. Appendix 6 illustrates the relationship between the activities in this manual and current national science education standards and benchmarks. There is also a general reference section and a glossary of technical terms.

This manual is intended for a diverse group of users. We have compiled the materials for use by the student and teacher as a team. All of the scientific information and exploration ideas are presented together so that the team has access to relevant information. Teacher Pages in each chapter are designed to help with the setup, management, and organization of the activities.

The levels of detail and complexity in this manual are balanced with the high school classroom as the midpoint audience. At the same time, we feel middle school and undergraduate classrooms will find these materials rich and accessible. Many of the activities and concepts are also suited for independent study by students or for use in informal education settings.

The address of the Fast Plants website is **www.fastplants.org**. Within it you can find more Fast Plants information and materials including current research, education activities for all levels, and links to related sites. It is also the communication gateway to the growing community of Fast Plant users including researchers, students, and teachers.

Question, investigate, and have fun. Life is out there. Explore it with Fast Plants!

Master Activity Calendar

The five chapters in this manual follow sequential stages in the Fast Plants life cycle. Within each chapter are individual activities that can guide students in the exploration of the topic. The chapters and their activities are as follows:

Chapter 1: Germination

Activity 1	Germination Observation
Activity 2	Imbibition: Swell Seeds
Activity 3	Dissecting the Seed
Activity 4	Germinating the Seeds: Which End is Up?

Chapter 2: Growth and Development

Activity 1	Tracking Variation
Activity 2	Hypocotyl Hypothesis

Chapter 3: Flowering

Activity 1	The First Flowers
Activity 2	The Flower Spiral
Activity 3	What's in a Flower?
Activity 4	Orientation of the Flower Parts
Activity 5	Go for the Glucose

Chapter 4: Pollination

Activity 1	Bees and Beesticks
Activity 2	It's Pollination Time
Activity 3	Pollen Germination

Chapter 5: Fertilization to Seeds

Activity 1	Embryogenesis—What, When, and Where?

The Master Activity Calendar on the following pages outlines how these activities fit together temporally.

The first week after planting the seeds, students can focus on germination activities and begin some growth and development activities.

During the second week, they could continue the growth and development activities and begin to get ready for pollination the following week.

The third week, when the plants are in full flower, the flowering and pollination activities can be carried out.

Activities in Fertilization to Seeds, (Chapter 5) are started in the third week, but are principally focused on the fourth, fifth, and sixth weeks.

With the exception of some of the activities the first week (the germination activities and the Hypocotyl Hypothesis), all activities can be carried out on the same set of plants. If following a set of plants through all chapters, we suggest that each student have two plants and some classroom "spares" be maintained in case of a "plant disaster"!

Master Activity Calendar

	MONDAY	TUESDAY	WEDNESDAY	THURSDAY	FRIDAY
0 das		Imbibition: Swell Seeds	Germination Observation Germinating the Seed: Which end is up?	Tracking Variation Hypocotyl Hypothesis	
7 das	Bees and Beesticks (some time this week)		Tracking Variation (cont.)		
14 das	Dissection (some time this week) Orientation of Flower Parts (some time this week) Go For the Glucose (some time this week)	It's Pollination Time	Tracking Variation (cont.) The First Flowers The Flower Spiral Pollen Germination		Embryogenesis: What, when and Where?

MONDAY	TUESDAY	WEDNESDAY	THURSDAY	FRIDAY
Embryogenesis: What, 21 das/6 dap	when and where	? (cont.) →	→	→
Embryogenesis: What, 28 das/13 dap	When and Where	? (cont.) →	→	→
Embryogenesis: What, 35 das/20 dap	→ When and Where			

KEY

- - - - - - - Germination Chapter
▭ ▭ ▭ ▭ ▭ Growth & Development Chapter
. . . . Flowering Chapter
o o o o o o o o o o Pollination Chapter
~~~~~~~~ Fertilization to Seeds Chapter

das = days after sowing, *dap* = days after pollinating
~~~~~~~~

Stages in the Life Cycle

Wisconsin Fast Plants are rapid-cycling brassicas, members of the cabbage and mustard family. Under proper growing conditions, the life cycle of 35–45 days will look approximately like this:

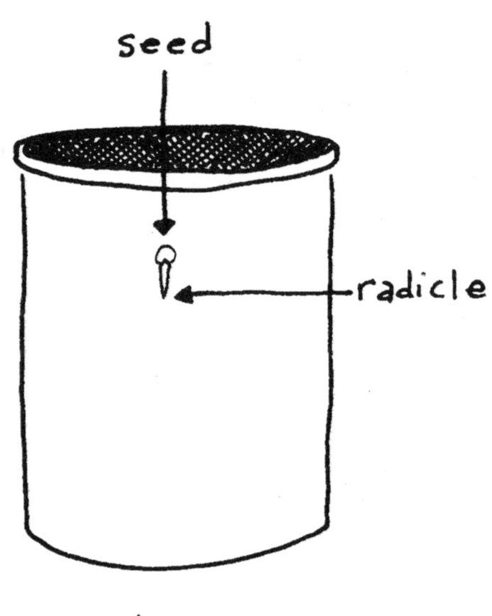

24 hours

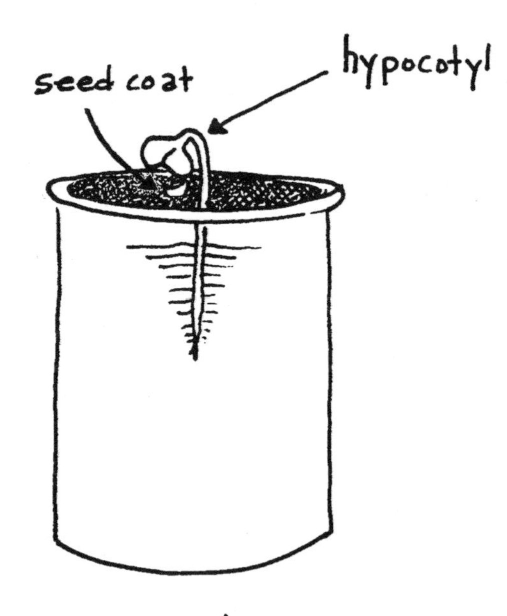

24 to 28 hours

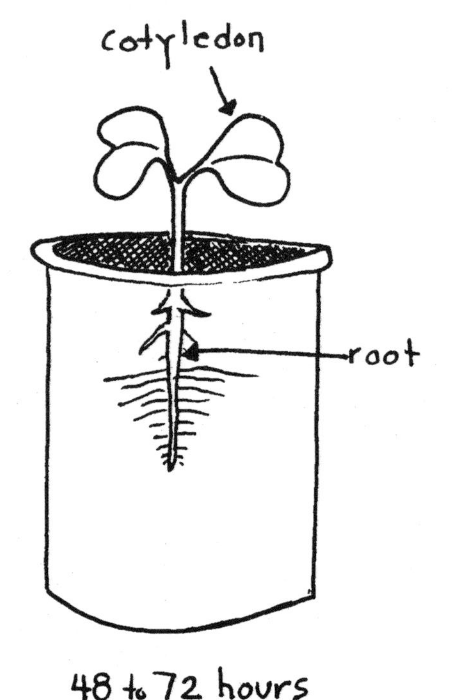

48 to 72 hours

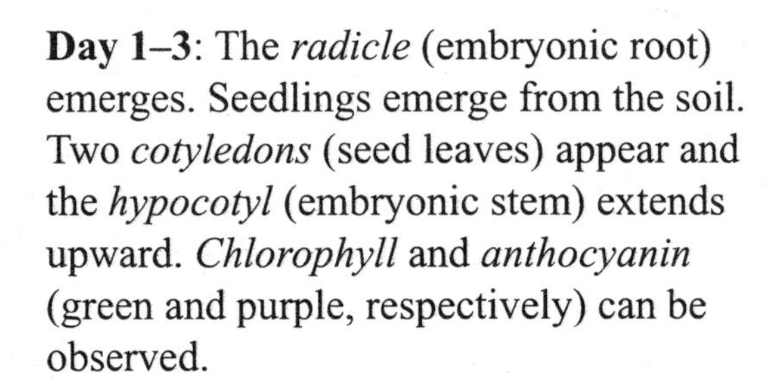

Day 1–3: The *radicle* (embryonic root) emerges. Seedlings emerge from the soil. Two *cotyledons* (seed leaves) appear and the *hypocotyl* (embryonic stem) extends upward. *Chlorophyll* and *anthocyanin* (green and purple, respectively) can be observed.

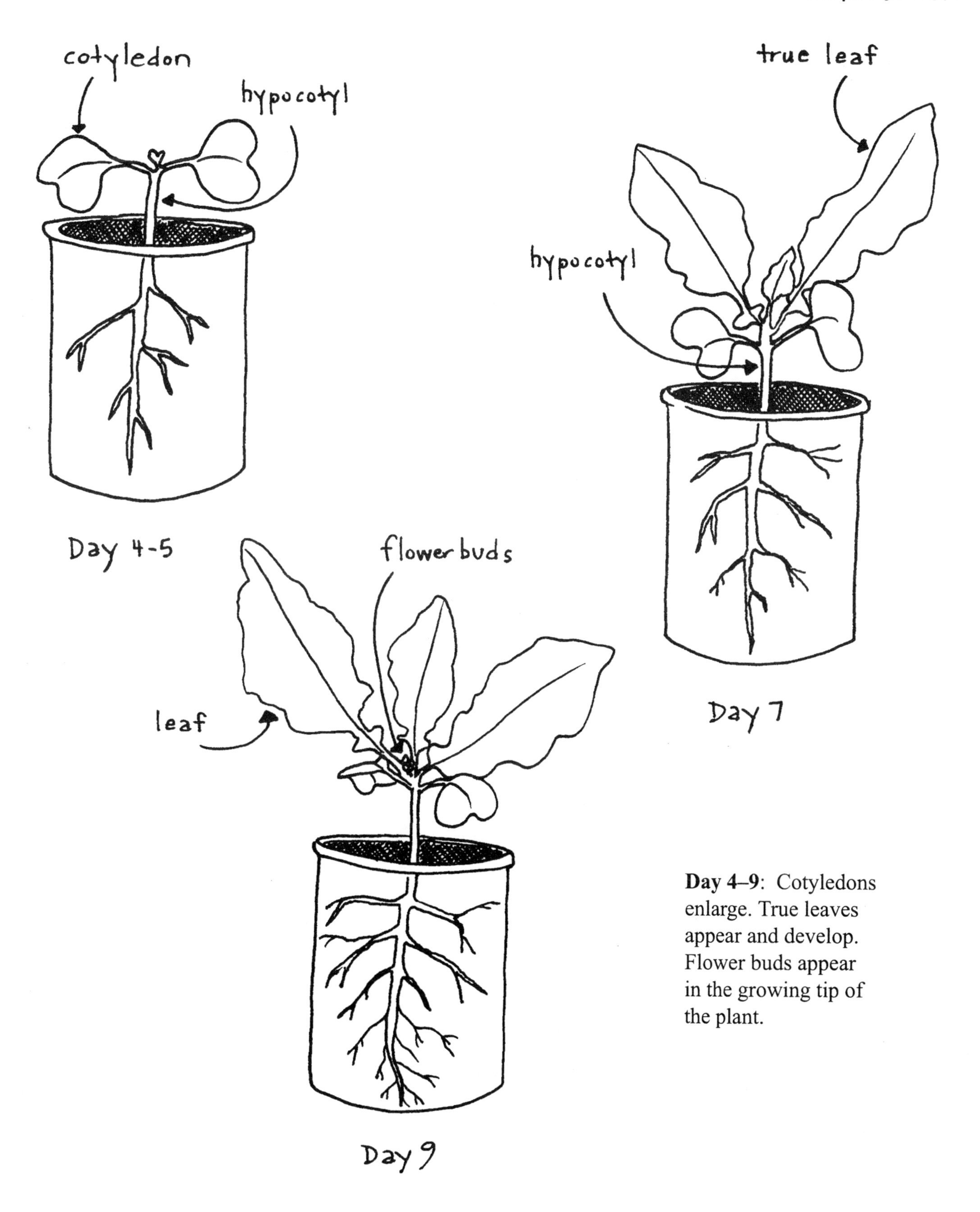

Day 4–9: Cotyledons enlarge. True leaves appear and develop. Flower buds appear in the growing tip of the plant.

Day 10–12: Stem elongates between the *nodes* (points of leaf attachment). Flower buds rise above the leaves. Leaves and flower buds continue to enlarge.

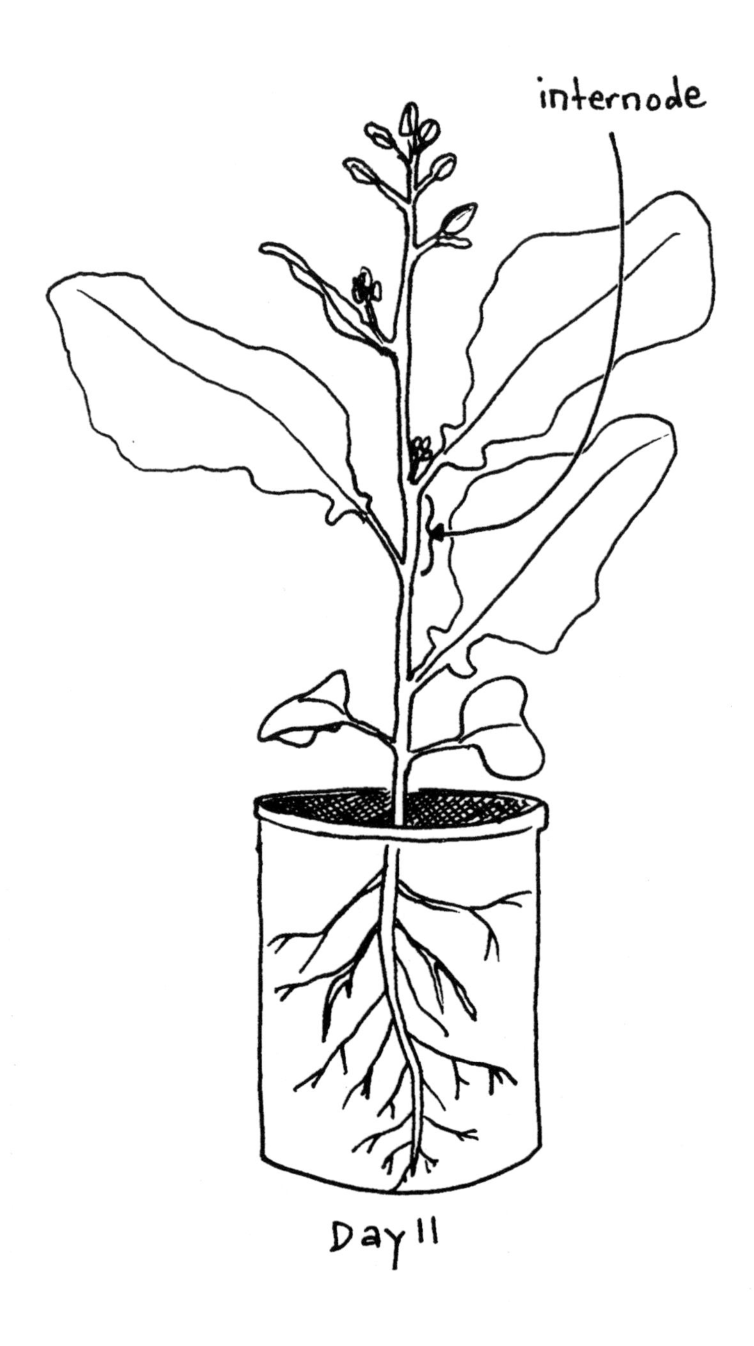

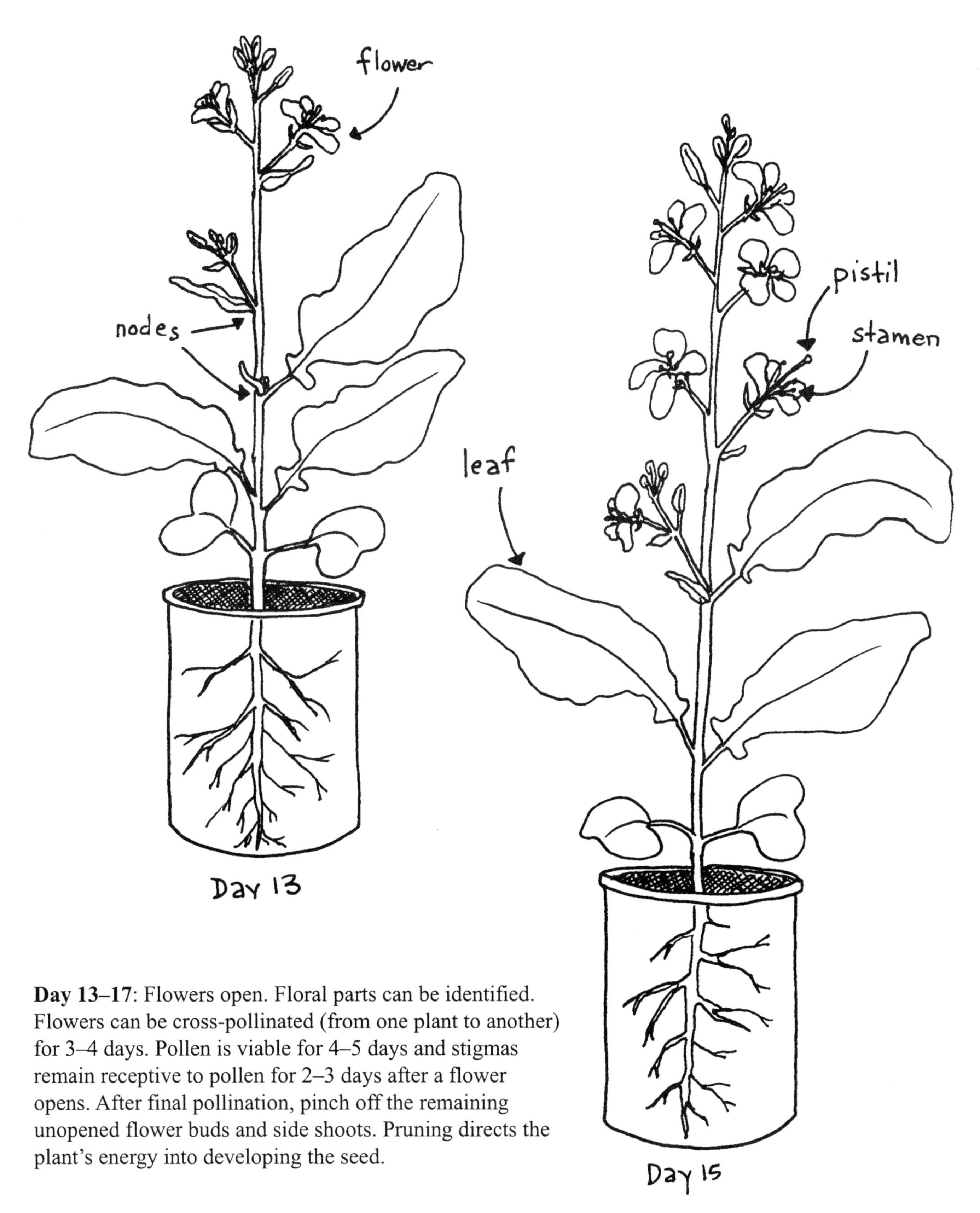

Day 13–17: Flowers open. Floral parts can be identified. Flowers can be cross-pollinated (from one plant to another) for 3–4 days. Pollen is viable for 4–5 days and stigmas remain receptive to pollen for 2–3 days after a flower opens. After final pollination, pinch off the remaining unopened flower buds and side shoots. Pruning directs the plant's energy into developing the seed.

Day 18–22: Petals drop from the pollinated flowers. Pods elongate and swell. Development of the seed and young plant has begun and will continue until approximately Day 36.

Day 18

Day 23–38: Seeds mature and ripen. Lower leaves yellow and dry. Twenty days after final pollination (about Day 38) plants should be removed from water.

Day 38–45: Plants dry down and pods turn yellow.

On **Day 45**, pods can be removed from dried plants. Seed can be harvested. The cycle is complete.

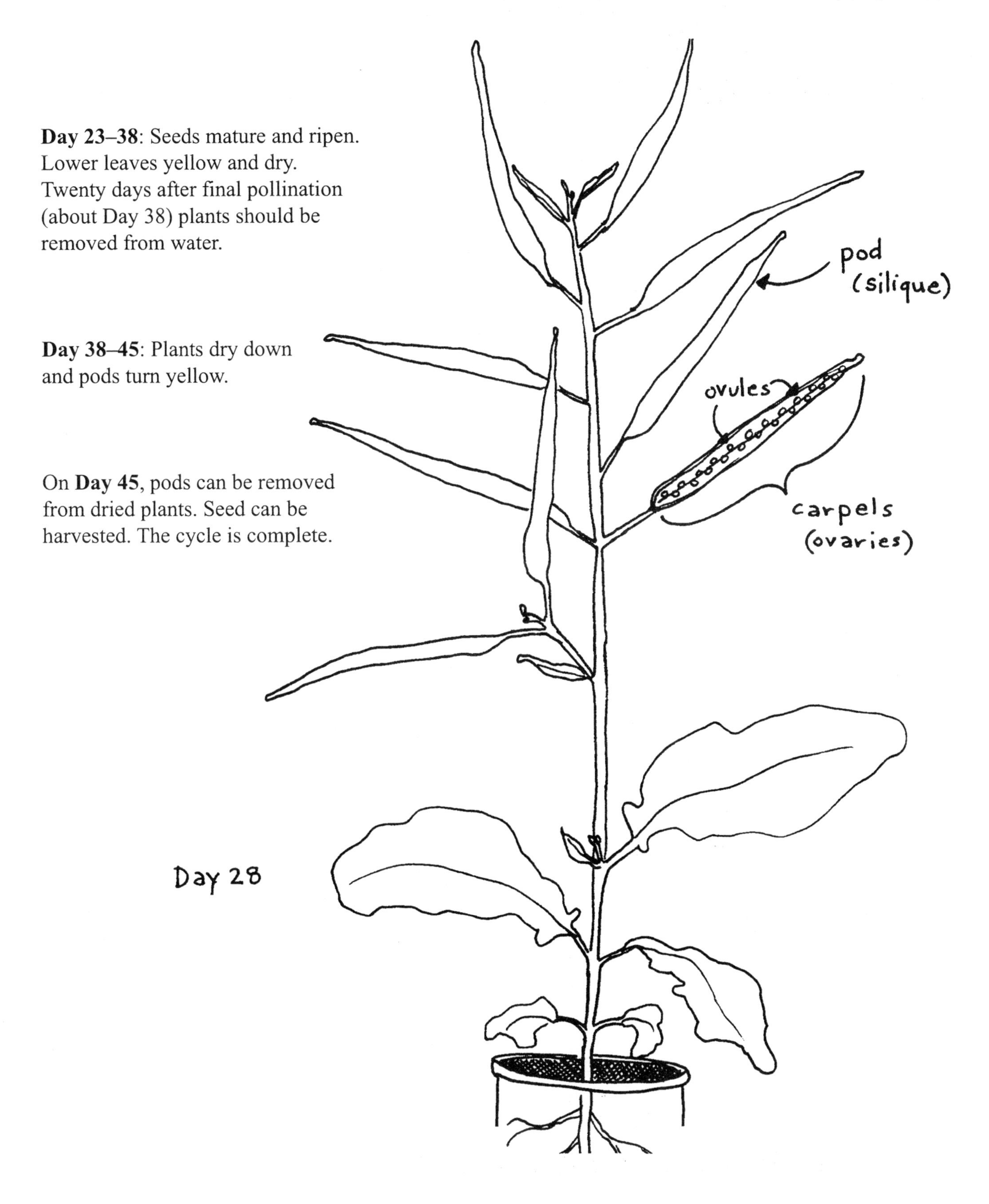

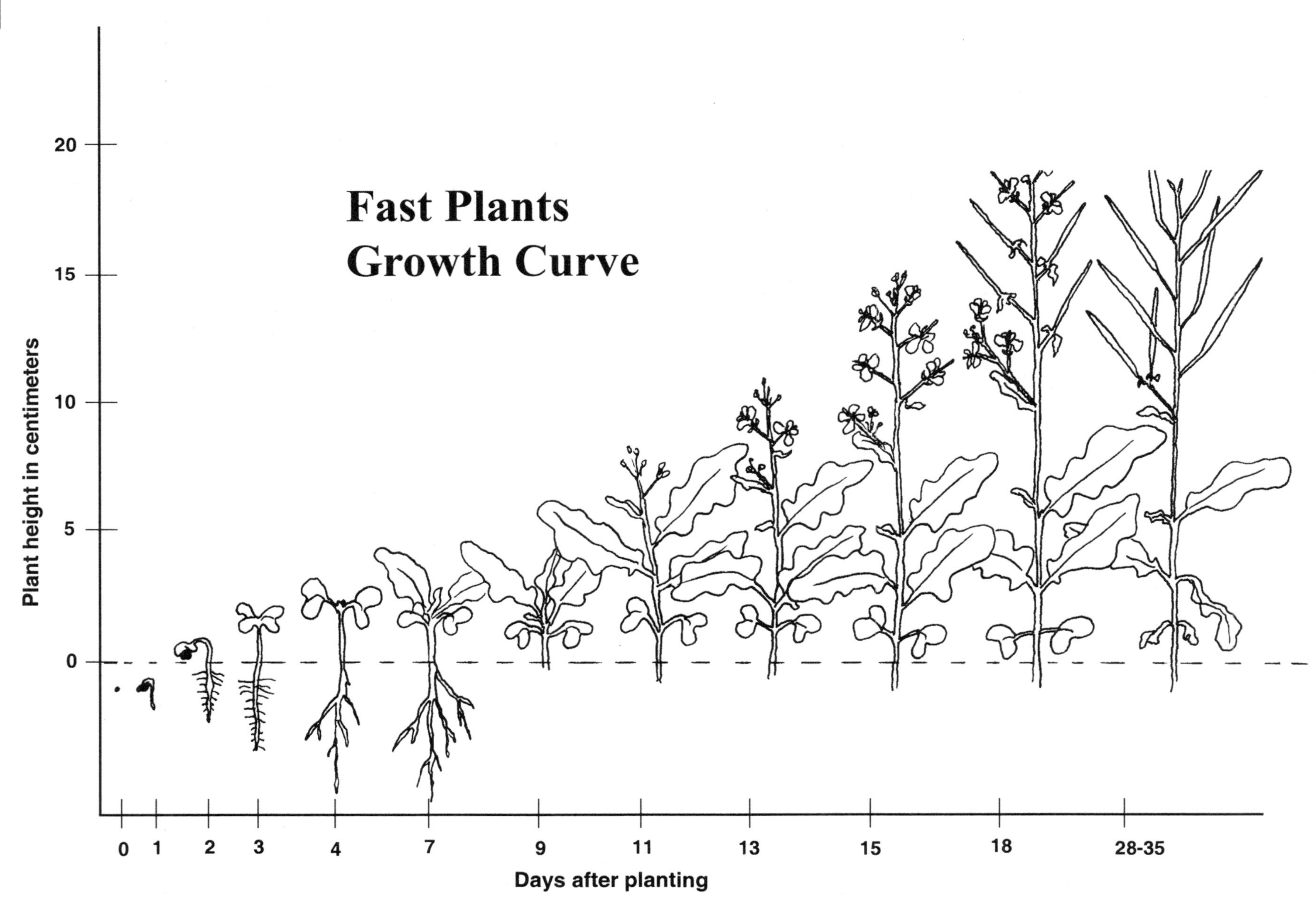
Fast Plants
Growth Curve
Plant height in centimeters
20
15
10
5
0
0
1
2
3
4
7
9
11
13
15
18
28-35
Days after planting

Grower's Calendar

Day of cycle	Maintenance
Preparation	Build/assemble light houses, wickpots** and reservoirs. Arrange all planting materials. See Appendix 1 page 113.
Day 1 (1 hour)	**Start life cycle** Plant, water from above, label, set wickpot on water mat with top of wickpot 5–10 cm from the lights.
Day 2–3* (5 min.)	Water from top with pipet. Cotyledons emerge.
Day 4–5* (15 min.)	Thin to 1 or 2 plant(s) per cell. Check the water level in the reservoir.
Day 6–11* (5 min./day)	**Check plants and reservoir level daily throughout the rest of the life cycle.** Observe growth and development.
Day 12* (30 min.)	Flower buds beginning to open.
Day 13–18* (15 min./day)	Pollinate for 2–3 consecutive days. On the last day of pollination, pinch off any remaining unopened buds.
Day 18–35*	Observe seed pod development. Embryos mature in 20 days.
Day 35–38*	Twenty days after the last pollination, remove plants from water mat. Allow plants to dry for 5 days.
Day 35–45* (30 min.)	Harvest seeds from dry pods. Clean up all equipment. Plant your harvested seeds or store them appropriately.

* Days to flowering will vary depending on environmental conditions in your classroom.
** "Wickpot" is a generic term for the growing container you may choose to use.

CHAPTER 1

GERMINATION

How does a seed become a plant?

BACKGROUND

Germination is the awakening of a seed from a resting state. This resting state represents a pause in growth of the embryo. The resumption of growth, or germination, involves the harnessing of energy stored within the seed. Germination requires at least water, oxygen, and a suitable temperature.

For many seeds, water is the "on" switch that initiates germination. As the dry seed imbibes or takes up water, the cellular cytoplasm hydrates. This hydration activates enzymes that then solubilize and loosen the structural fabric of cell walls. As additional water is absorbed, the seed's cells enlarge and the seed coat cracks. A *radicle* (embryonic root) emerges from the seed. Rapid development of the fine root hair cells vastly increases the surface area of the root, facilitating the uptake of more water. In Fast Plants and many other dicots, this uptake of water drives the elongation of the *hypocotyl* (embryonic stem), which pushes the *cotyledons* (seed leaves) and shoot meristem upward through the soil.

Food for Thought...

SLEEPING BEAUTY

Some seeds can remain dormant and capable of germinating for decades, even centuries. With proper environmental conditions, these emissaries from long ago can "launch," grow, and reproduce. One dramatic example involves the ancient seeds of *Canna compacta*. These small seeds had been inserted by people into young, soft walnut fruits so that when the fruits healed and the shells hardened, the seeds were entirely enclosed in the shells. Subsequently the nut and seeds would dry out, creating rattles used for ornamentation. Recent excavation of a tomb in Argentina uncovered these rattles containing still-viable *C. compacta* seeds. The hardened shells enclosing the seeds was carbon dated at 600 years old.

What mechanisms allow some seeds to remain viable for centuries while others lose viability in a few years? Plant scientists are very interested in understanding factors that maintain seed viability to better understand how to increase seed longevity, maintain agricultural varieties of crop plants, and preserve seeds of wild and endangered species. ⊠

GERMINATION

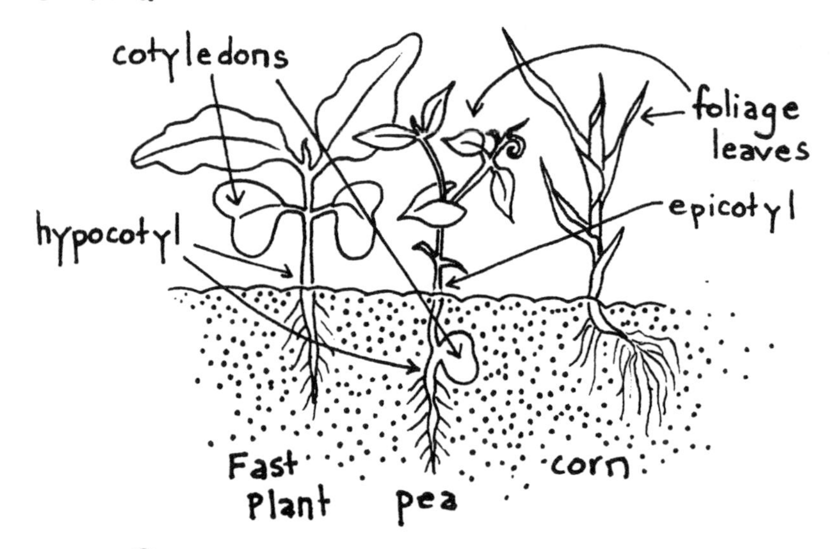

FROM LIPIDS to LIGHT

Food for Thought...

In dicotyledonous plants, up to 90% of a mature seed is composed of cotyledons that are packed with fuel molecules (starches, lipids, and proteins). Accompanying these energy-storing compounds is a parallel set of subcellular machinery (enzymes and organelles) poised to rapidly harness this fuel during the early stages of germination. In Fast Plants and related brassicas this energy source is mainly stored as oils. These oils are located in organelles called oleosomes. Many of the enzymes required to metabolize the lipids during germination are located in yet another organelle, the glyoxysome.

During the later stages of germination an extraordinary transformation takes place in the cells of these cotyledons. In a period of one to two days the cotyledons emerge above the ground and change from storage to photosynthetic tissues. The oleosomes and glyoxysomes are in part replaced by a new set of organelles including chloroplasts and peroxisomes. Chloroplasts contain chlorophyll and many of the enzymes associated with photosynthesis.

BACKGROUND, continued

Not all plants germinate in such a fashion. Cotyledons from pea plants remain below the ground. The shoot tip is lifted out of the soil by the elongation of the *epicotyl* (the embryonic stem above the cotyledons). Monocots, such as grasses, push the *coleoptile* (a protective sheath) from the seed upward through the soil. The shoot tip then extends through the coleoptile and out of the soil.

In Fast Plant germination, cotyledons emerge from the soil, expand, cast off the protective seed coat, turn green, and become photosynthetically active. At this point the plant becomes independent of its stored reserves and dependent on the energy of light. Launch has been successful! All of these events happen on Days 1, 2, and 3 of the Fast Plant life cycle.

We still do not completely understand seed germination and scientists are very

Cotyledon Cells

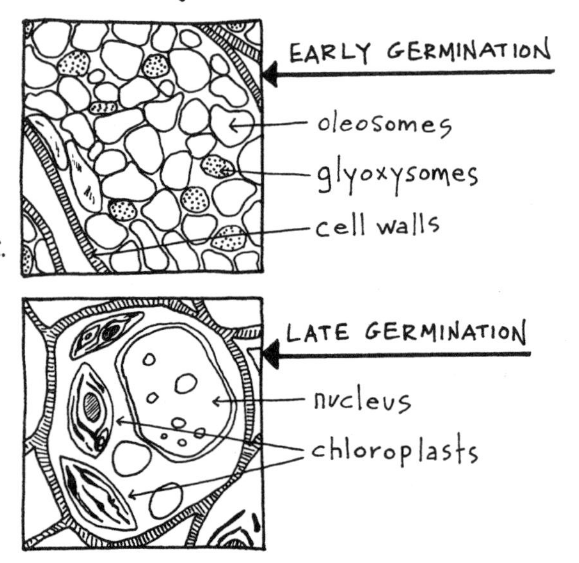

interested in learning more about it. Because germination holds so many unanswered questions, it can be an excellent topic for investigation.

Activities

Activity 1
The Germination Observation
When placed under favorable conditions, Fast Plant seeds will germinate quickly. Germination involves the expanding and propelling of two growing points of the seed outward (upward and downward if guided properly). The following activity provides an easy way to observe and carefully record the events of germination.

Variations in the physical environment such as temperature and light will alter the rate of germination and appearance of Fast Plant seedlings. Use this activity to investigate the interactions between the environment and germination.

Food for Thought...

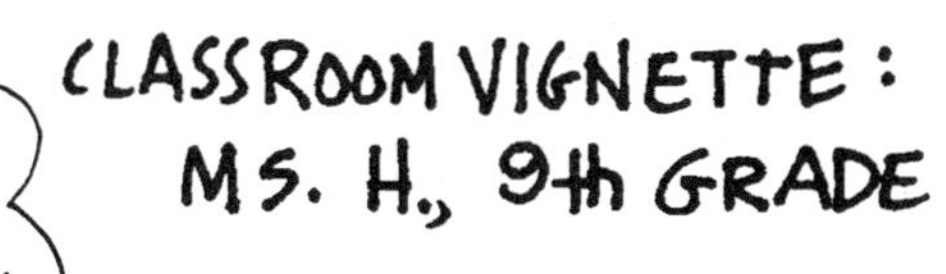

The first day of Biology I at Medfield High School .... Ms. H starts with the question, "What is biology?" Students suggest definitions such as "the study of life," "looking at living things," and "studying animals" as Ms. H. records their responses on the board.

As life seems to be a central theme in their answers, Ms. H. then asks, "What is life?" Again a list is generated on the board as the students begin to define a living thing: it breathes, it moves, it is made of cells, it makes noise, and it needs water.

"If biology deals with living things," suggests Ms. H., "we need to get a better idea of what 'living' really means. As a matter of fact, there is a living organism right next to you at your lab bench."

"Huh?" A few startled students begin to glance about with some trepidation.

"No kidding," Ms. H. continues. "There is a living organism within one meter of you!" ...

Prior to students entering the room, Ms. H. has taped a Fast Plant seed near each student station...

Later, as students protest that a seed shouldn't count as a living thing, Ms. H. returns their question, "How could you determine if a seed is indeed living?"... ⊠

See Appendix 5 (page 139) for the full vignette.

ACTIVITIES, continued

Procedure

1. From a paper towel or a piece of filter paper, cut a circle 8.5 cm in diameter to fit in the cover (larger half) of a petri dish. With a pencil, label the bottom of the paper circle with your name, the date and the time.

2. Place a transparency-plastic Ruler Disk in the cover of the germination petri dish; place the paper circle on top. (The ruler will show through the paper circle once it is wet.)

3. Moisten the paper circle with an eye dropper.

4. Place five Fast Plant seeds on the paper circle along the middle dark line on the ruler and cover with the bottom (smaller half) of the petri dish.

5. Place the petri dish at a steep angle (80°–90°F) in shallow water in a tray so that the bottom two centimeters of the paper are below the water's surface.

6. Set the experiment in a warm location (optimum temperature: 65–80°F). Check the water level each day to be sure the paper circle stays wet.
   ↪ On the Germination Observation individual data sheet (page 36) record the day, time, and initial environmental conditions for the experiment.

7. Over the next 3–4 days observe the germinating seed and seedlings using a magnifying lens. (See Appendix 3, page 133 for instructions on making an inexpensive film can hand lens.)
   ↪ Measure and record the growth of the roots and shoots. Sketch the germinating seeds and young plants

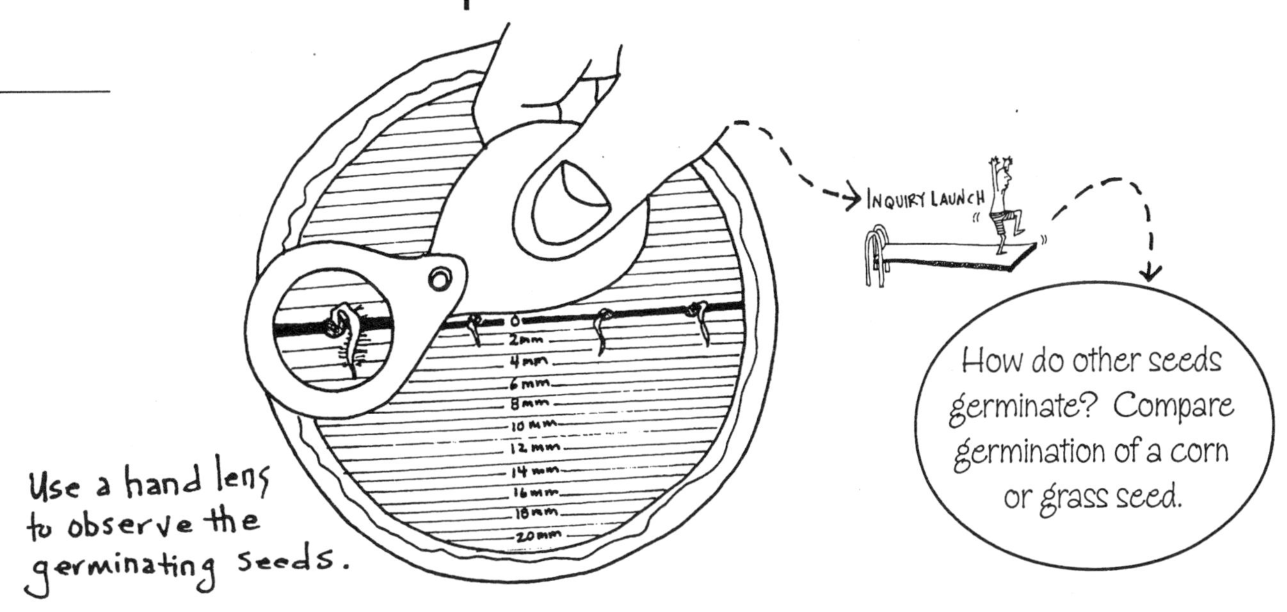

using a hand lens/magnifying glass. A Scale Strip (Appendix 4, page 135) will help in measurement, sketching under magnification, and drawing to scale. Record all data on the data sheet.

8. Growth that orients to conform with the direction of gravity is referred to as positive gravitropism while growth that orients against the force of gravity is negative gravitropism.

↪ Note positive and negative gravitropism on your data sheet.

↪ Graph the combined length of the Fast Plant roots and hypocotyl (dependent variable = y-axis) over time (independent variable = x-axis) on the "Germination Observation" sketch sheet and data analysis sheet (page 37).

?? Would your graph look different if you plotted only root or hypocotyl growth?

?? Does a germinating seed "know" which way to grow?

?? Consider compiling a class data set. Does a larger sample size affect your conclusions?

THE WAKE-UP CALL

Food for Thought...

Seed dormancy is a key to survival for many plants. It provides some insurance that germination will occur at a time and place advantageous to the seedling. Breaking dormancy can require certain environmental conditions, such as a heavy rainfall for some desert plants, a prolonged cold period for many plants in a harsh winter climate, or exposure to stomach acids for seeds dispersed by animal consumption.

In order to break dormancy for horticultural use, many wild seeds must be treated in a way that mimics natural conditions. Some seeds must be exposed to weeks or months of cold (and sometimes moist) conditions in a process called *stratification*. Other seeds must be physically altered through *scarification*. In this process, the seed coat barrier is altered through use of knives, files, sandpaper, alcohol, solvents, or acids.

ACTIVITIES, continued

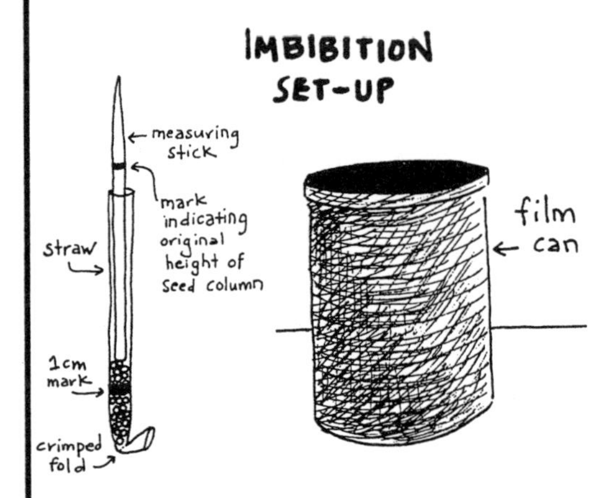

Activity 2
Imbibition: Swell Seeds
Many seeds are awakened from their dormant stage by moisture. The initial stage of germination when the seed takes up water is called *imbibition.* What actually happens during imbibition? What controls the process?

By setting up an experiment in a film can you can carefully observe the imbibition process. Take your experiment home so you can observe the responses of the seeds periodically.

Procedure

1. Add 2–3 ml of water to a film can.

2. Take a 4.5 cm long piece of a narrow (2–3 mm) plastic straw, bend the bottom 5 mm sharply to form a fold in the straw and bite the fold between your front teeth to crimp. With an ultra-fine point permanent marking pen and ruler, place a mark on the straw 1 cm above the crimped fold.

3. While counting the number, add seeds to the top of the straw until they have reached the 1 cm mark.

4. Insert the toothpick into the straw until it rests on the top of the seed column. Use the pen to mark the toothpick where it emerges from the straw. This mark will help you determine the baseline measurement for the seed column height and will aid in detecting how the height of the seed column changes throughout the experiment.

5. With a pipette, add water to the top of the straw until it begins to drip from the bottom of the folded portion.

6. Place the straw in the film can so the base of the straw is under the water. Put the cap on the can.

7. Wait 10–30 minutes, open the film can and insert the toothpick into the straw until it rests on the seed column again. Mark the toothpick again where it emerges from the straw.

?? Is the environment within the can changing? Record your observations.

8. For a period of 24–48 hours, measure the height of the seed column in the straw every few hours by marking the toothpick.

↪ Note the height and time of each measurement on the Imbibition: Swell Seeds data sheet (page 38).

↪ Graph the height of the seed column (dependent = y-axis) over time (independent = x-axis) on the data sheet.

9. After 24–48 hours, use the toothpick to push the column of seeds onto a paper sheet. Observe the appearance of various seeds in different places along the column and record your observations.

Food for Thought...

TASTY BRASSICAS

Fast Plants, like all brassicas, contain chemicals called glucosinolates. When activated by the saliva in your mouth, glucosinolates give the typical flavors found in all brassicas (such as broccoli, cabbage, Brussels sprouts, and radish).

These biologically active chemicals may be beneficial in that they may repel certain insects while attracting others. The strong flavor helps to discourage deer and rabbits from eating brassica crops in the fields. In mammals, some glucosinolates have been found to help detoxify cancer-causing nitrosamine chemicals in the liver. Chew a Fast Plant seed and see how it tastes—Is it spicy? Compare it to a mustard seed.

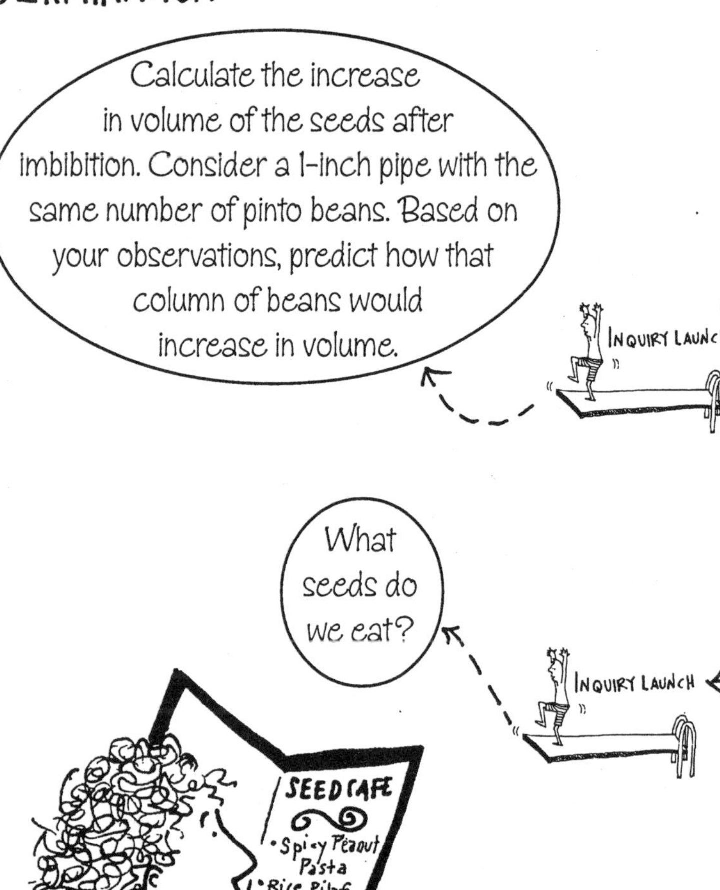

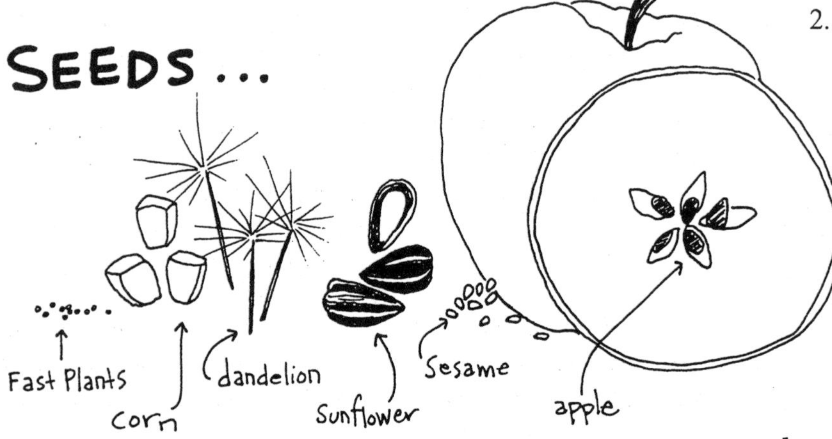

ACTIVITIES, continued

?? Did all of the seeds swell? Did any germinate? What variation in the environmental conditions in the straw might explain your observations? Consider the environmental factors necessary for germination.

Activity 3
Dissecting the Seed

At first glance, many seeds seem to have no identifying features. But seeds are far from homogeneous solid spheres. In Fast Plants, for instance, seeds are internally packaged with the embryonic plant and the food-storing cotyledons. Clues to the seed's interior are visible on the external seed coat.

Procedure

1. Place a pinto bean and a brassica seed on a piece of double-sided sticky tape on a Scale Strip (Appendix 4, page 135). Roll the pinto bean around with a needle or pencil point and observe its shape and features. A distinctive oval light area on the seed coat will be observed. This is the *hilum*, the scar where the developing seed was attached through the *funiculus* (stalk of the ovule) to the maternal tissue of the ovary. If the hilum is facing you, this is the front of your seed.

2. Now roll the brassica seed around on the tape. With the aid of a magnifier, find the hilum, a darker circular area with a small lighter area within it.

3. With both seeds still on the sticky tape of the Scale Strip, place the strip under a dissecting microscope or lens with a magnification between 10X and 40X. At the higher magnification, not all of the pinto bean will fit into the field of view.

You will observe a minute hole, the *micropyle*, in a depression at one end of the hilum and opposite the end with two small raised pear-shaped structures. The micropyle is the hole in the ovule integument through which the pollen tube passed on its way to fertilization of the egg and polar nuclei. The micropyle is also a weak area in the seed coat, or testa, that splits under pressure from the emerging root tip.

↪ On the Seed Dissection sketch sheet (page 39), draw to scale the details of the hilum area, observing the location of the micropyle for both the pinto and brassica seed. In brassica seeds, the micropyle is less conspicuous than in the bean, but appears as a tiny raised area adjacent to the darkened circular area of funicular attachment.

3. Soak a pinto bean (4–12 hours) and a brassica seed (1–4 hours) in water to soften their seed coats. Dry off the excess water and place them on a Scale Strip under a dissecting scope.

4. Under magnification, examine the front views of the soaked seeds, comparing them with the dry seeds.

?? Has anything changed? Can the hilum and micropyle still be seen?

5. Keep the micropyle of the soaked pinto bean in view. With a sharp dissection needle cut through the testa around the hilum, peeling back the seed coat to expose the white or pale cream embryonic plant. As this is done, you will see the rounded tip of the embryonic root pointing towards the micropyle.

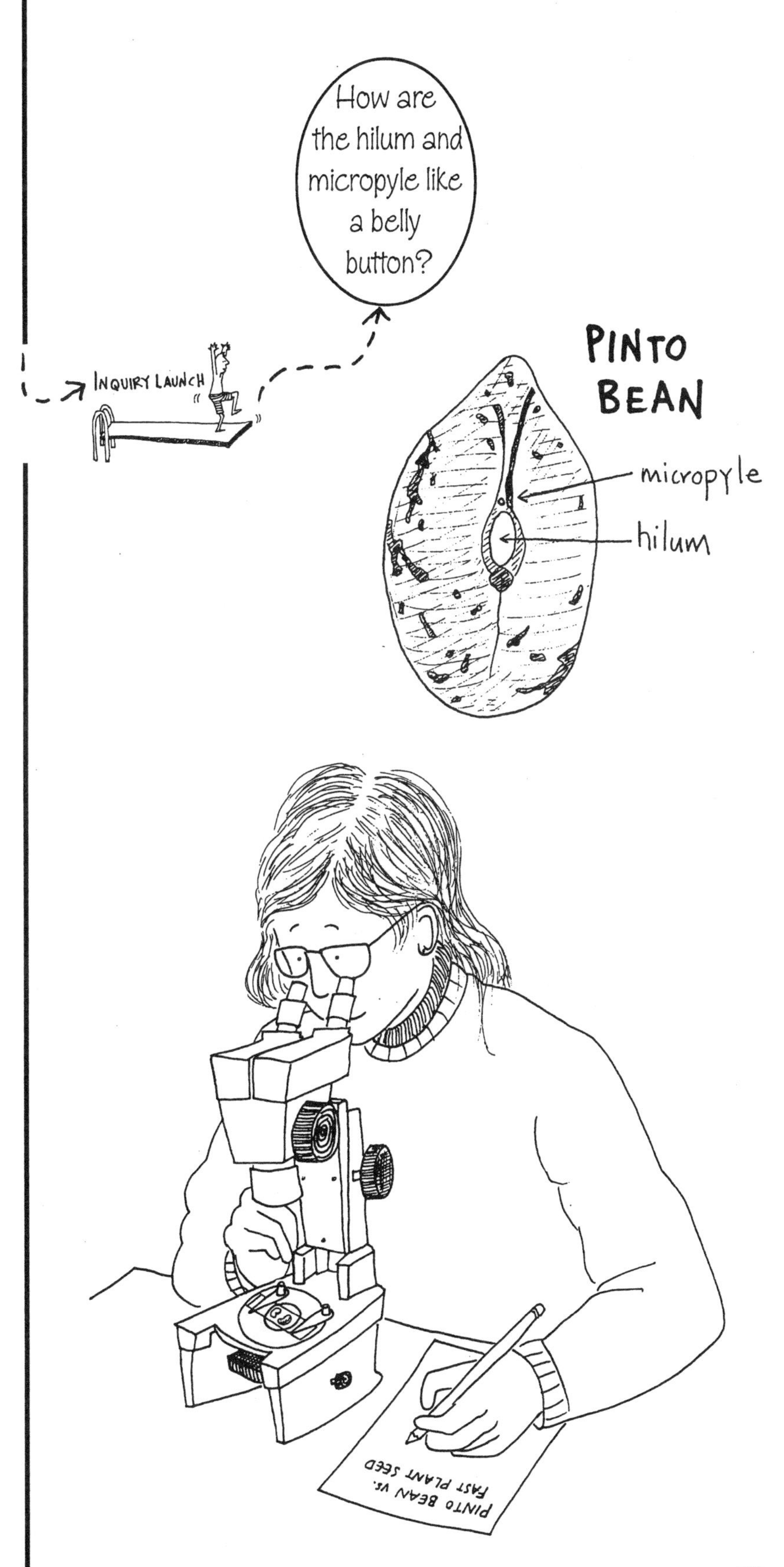

GOOD THINGS COME IN SMALL PACKAGES

Food for Thought...

About 70% of all food for humans comes directly from seeds and a large proportion of the remainder is derived from animals that feed on seeds (Bewley and Black, 1994). Seeds store extra energy reserves in the form of carbohydrates, oils or proteins. These food reserves support early seedling growth before the plant is able to photosynthesize.

Scientists are very interested in learning more about the nutritional content of wild species and wild progenitors of our cultivated crops, as these species may represent genetic diversity that could provide new sources of nutrition in the improvement of domesticated crops. ⊠

ACTIVITIES, continued

?? How is the embryonic plant packaged in the seed?

↪ On your data sheet, draw a front view of the embryo in the orientation with root tip pointing down.

?? What else do you see inside the seed? What function(s) are associated with the remainder of the contents of the seed?

Activity 4
Germinating the Seed: Which End is Up?

If the embryonic plant is packaged in the seed in a consistent manner, what happens if a seed germinates in a different orientation? Does the root always grow down? Construct a bottle cap seed germinator to explore this question. Like the imbibition investigation, you can take your experiment home and observe it every

The Food Reserves of Some Important Species*

	Average Percent Composition			Major
	Protein	Oil	Carbohydrate	Storage Organ
Cereals (monocots)				
Barley	12	3	76	Endosperm
Dent corn (maize)	10	5	80	Endosperm
Oats	13	8	66	Endosperm
Rye	12	2	76	Endosperm
Wheat	12	2	75	Endosperm
Legumes (dicots)				
Broad bean	23	1	56	Cotyledons
Garden pea	25	6	52	Cotyledons
Peanut	31	48	12	Cotyledons
Soybean	37	17	26	Cotyledons
Other				
Brassica	21	48	19	Cotyledons
Castor bean	18	64	Negligible	Endosperm
Oil palm	9	49	28	Endosperm
Pine	35	48	6	Megagametophyte

*modified after Bewley and Black (1994).

few hours. Be sure to keep it in the vertical position with the proper north-south-east-west orientation.

Procedure

1. Cut two layers of absorbent kitchen paper towel into 26 mm circles (easily done by tracing a U.S. quarter). These will fit neatly into the bottom of a soda bottle cap.

2. Place the towel in the cap and moisten it with water. Pour off any excess.

3. Orient each of the seeds in one of four positions: top, bottom, right and left on the moist towel surface, making sure that the brown micropyle area (spot) is pointing toward the center of the cap, as shown in the illustration.

4. Make a mark on the bottle cap that indicates up.

5. Cover the open cap with plastic wrap and secure the wrap with a rubber band. Trim off excess wrap with scissors.

6. Position the "bottle cap seed germinator" upright by standing it in a second bottle cap.

↪ On Circle 1 of the Germination Chamber sketch sheet (page 40) draw the seeds in the bottle cap germinator, including the orientation of the brown micropylar area toward the center of the circle.

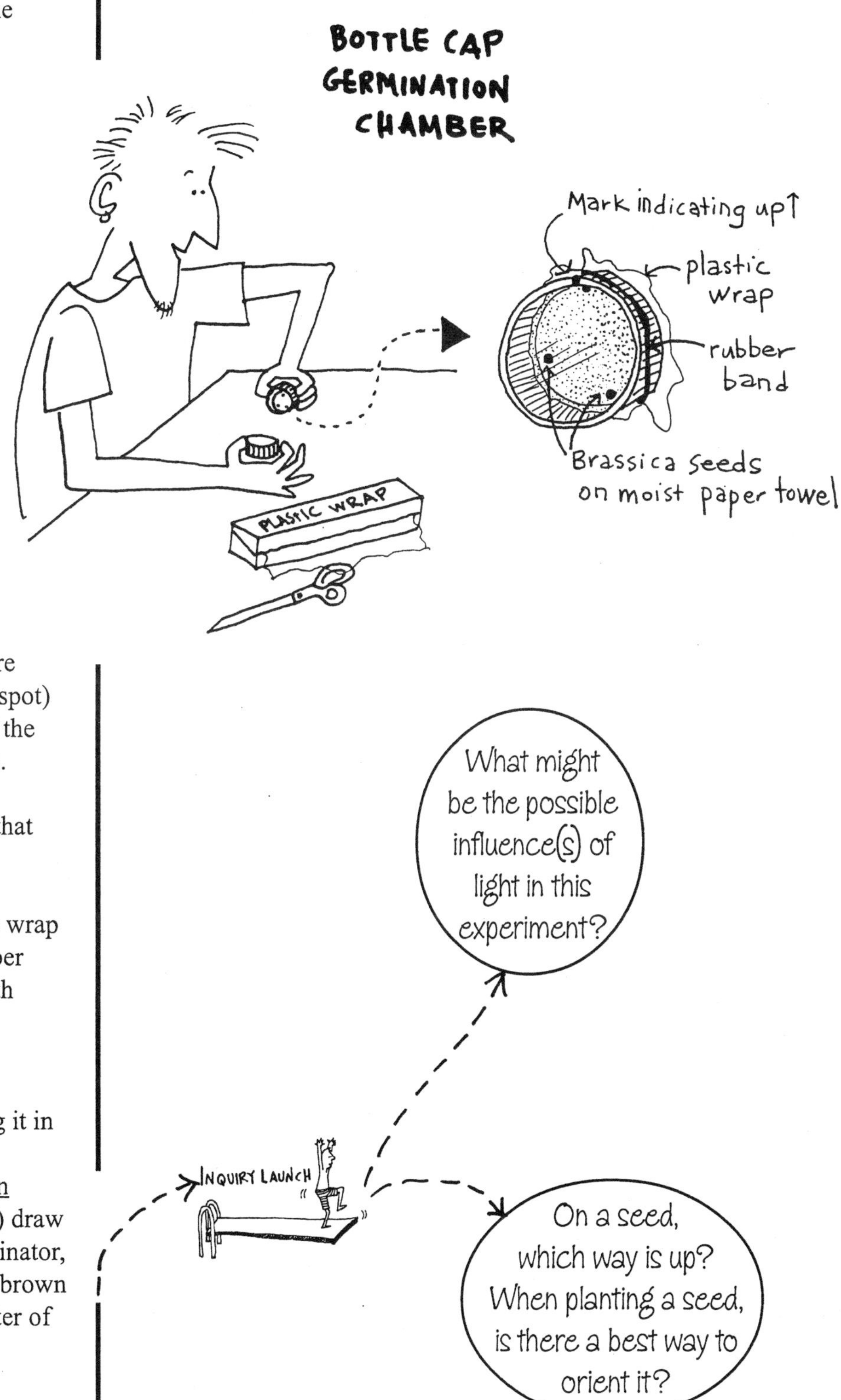

ACTIVITIES, continued

↪ When the roots begin to emerge (approximately 8–12 hours), record the direction of the emerging root from each seed with a second drawing in Circle 2 of the sketch sheet. You may wish to use a hand lens.

?? Based on your observations, is there an "up" end to the seed? If there is, which end is "up"?

7. Continue to observe the germinating seeds, being sure that the paper towel is kept moist. Note the appearance of the fine, fuzzy root hairs and the extension of the hypocotyl.

↪ After 24–48 hours, make a third drawing depicting the direction of the roots and hypocotyl and illustrating the root hairs, cotyledons and seed coat (Circle 3).

?? How do plants "know" which way to grow?

Additional Reading

J. D. Bewley, “Seed Germination and Dormancy,” *Plant Cell* **9**:1055–1066, 1997. Details the processes associated with germination and describes some possible mechanisms for dormancy.

J. D. Bewley & M. Black, *Seeds: Physiology of Development and Germination* (second edition), Plenum Press, New York, 1994.

A. N. C. Huang, R. Trelease & T. S. Moore, Jr., *Plant Peroxisomes*, Academic Press, New York, 1983. Provides detailed information on glyoxosomes and peroxisomes.

C. M. Karssen, “Hormonal Regulation of Seed Development, Dormancy, and Germination Studied by Genetic Control,” in *Seed Development and Germination* (J. Kigel & G. Galili, editors), pp. 333–350, Marcel Dekker, Inc., New York, 1995.

A. M. Mayer & Y. Shain, “Control of Seed Germination,” *Annual Review of Plant Physiology* **25**:167–193, 1974.

Overview

These activities will provide students with insight into the various stages of germination. In Activity 1, students observe germination and record growth rates of roots and shoots. In Activity 2, students explore and measure imbibition, which triggers the germination process in seeds. Activity 3 gives the students an opportunity to become acquainted with the anatomy and biology of seeds and begin to answer the question posed in Activity 4. In this final activity, students observe and record direction of root emergence from the seed in order to get a handle on whether there is an "up" and "down" to seeds relative to their anatomy and germination.

Objectives

By participating in this unit, students will understand that:

- germination is the beginning of growth of a new plant from a seed;
- a seed is a dry, dormant, embryonic plant complete with a reserve of stored energy to keep it alive and sustain germination;
- moisture, temperature, and oxygen regulate the germination of seeds;
- water is taken up by seeds in order to initiate germination;
- germination commonly involves emergence of an embryonic root followed by an embryonic stem that, in the case of Fast Plants, pushes the cotyledons and shoot meristem through the soil;
- the external anatomy of seeds is associated with certain features of the internal anatomy of seeds;
- plant roots display positive gravitropism while the hypocotyl and shoots are negatively gravitropic; and
- emergence of the seedling and production of chlorophyll in the cotyledon prepares the plant for growth in the presence of sunlight.

Time Required

For All Activities

- One to two periods can be spent discussing results as well as graphing and statistically analyzing data.

Activity 1:
Germination Observation

- **Day 1**: one 50-minute class period on first day to set up germination plates and record initial data.
- **Days 2–4**: 10–20 minutes on each day to observe and record data.

Activity 2:
Imbibition: Swell Seeds

- **Day 1**: one 50-minute class period on first day to set up experiment and record initial data.
- Five minutes for observation and data recording every few hours for 24–48 hours.

Activity 3:
Dissecting the Seed

- One 50-minute class period.

Activity 4:
Germinating the Seed: Which End is Up?

- **Day 1**: one 50-minute class period.
- **Day 2**: 10–20 minutes to observe and record data.
- **Day 3 or 4**: 15–25 minutes to observe and record data, reorient germinator, and record a new prediction.
- **Day 5**: 10–20 minutes to observe and record data.

Materials

Each student will need:

Activity 1

- 1 petri dish
- photocopy of the ruler disk on p. 41 onto a transparency sheet, cut out
- 5 Fast Plants seeds
- paper towels
- eye dropper
- hand lens
- shallow tray or bottom from a 2-liter soda bottle

Activity 2

- Scale Strips (Appendix 4, page 135) or short metric ruler with gradations in mm
- film can
- scissors
- 4.5-cm piece of 2–3 mm diameter transparent cocktail straw
- toothpick
- ultra-fine point permanent marking pen
- 15–30 Fast Plant seeds (see alternative in Classroom Management Tips)

Activity 3

- Scale Strips (Appendix 4, page 135)
- 2 Fast Plant seeds, one dry and one presoaked in water for 1 to 4 hours
- 2 pinto bean seeds, one dry and one presoaked in water for 4 to 12 hours
- 5X–10X hand lens or dissecting microscope
- dissecting needle

Activity 4

- two soda bottle caps (film can lids will also work)
- kitchen plastic wrap
- paper toweling
- 4 Fast Plant seeds
- forceps for handling seed
- hand lens
- rubber band

Classroom Management Tips

- Set up the first activity, *Germination Observation*, in the morning so students can observe the seed again just before leaving school. If started on Monday, the investigation can be completed by Friday.

- Germinated seedlings can be carefully transplanted into wickpots at Day 5 and grown to maturity. This may slow the developmental cycle by a few days.

- Start the activity, *Germinating the Seed: Which End is Up?*, on a Monday, so that germination will occur during the week when students can make observations. Alternatively, students may set it up and observe it at home.

- The activity, *Imbibition: Swell Seeds*, uses quite a lot of seeds (15–30 per student). If your students are producing lots of Fast Plant seeds, these can be used. You can also use other seeds such as turnip, Chinese cabbage, radish, cabbage, mustard, etc., obtained from local garden or grocery stores or mail order seed supply companies.

FAST PLANTS: GERMINATION ACTIVITIES CALENDAR

MONDAY	TUESDAY	WEDNESDAY	THURSDAY	FRIDAY
GERMINATION OBSERVATION: put seeds on petri plate	→	→ observe and	Record →	→
IMBIBITION: put seeds in straw	→ observe and	Record: EXAMINE SEEDS		
GERMINATION CHAMBER: place 4 seeds in film can	→ observe and	Record: CHANGE ORIENTATION →	→ observe and Record	

Germination Observation

Individual Data Sheet

How do germinating seedlings grow?

Seedling Length* in Petri Dish

Date	Time	Hours after starting exp.	Plant #1 length	Plant #2 length	Plant #3 length	Plant #4 length	Plant #5 length

Seedling Length—Analysis of Individual Data

Hours after starting exp.	Average	Range	Standard deviation

Initial Environmental Conditions

Temperature ___________

Light (circle one)

lightbank/lighthouse

room light

window

(facing which direction?_____)

other ___________

* Seedling length = shoot length + root length

Germination Observation

Sketch Sheet and Data Analysis

Sketch of Germinating Seedlings

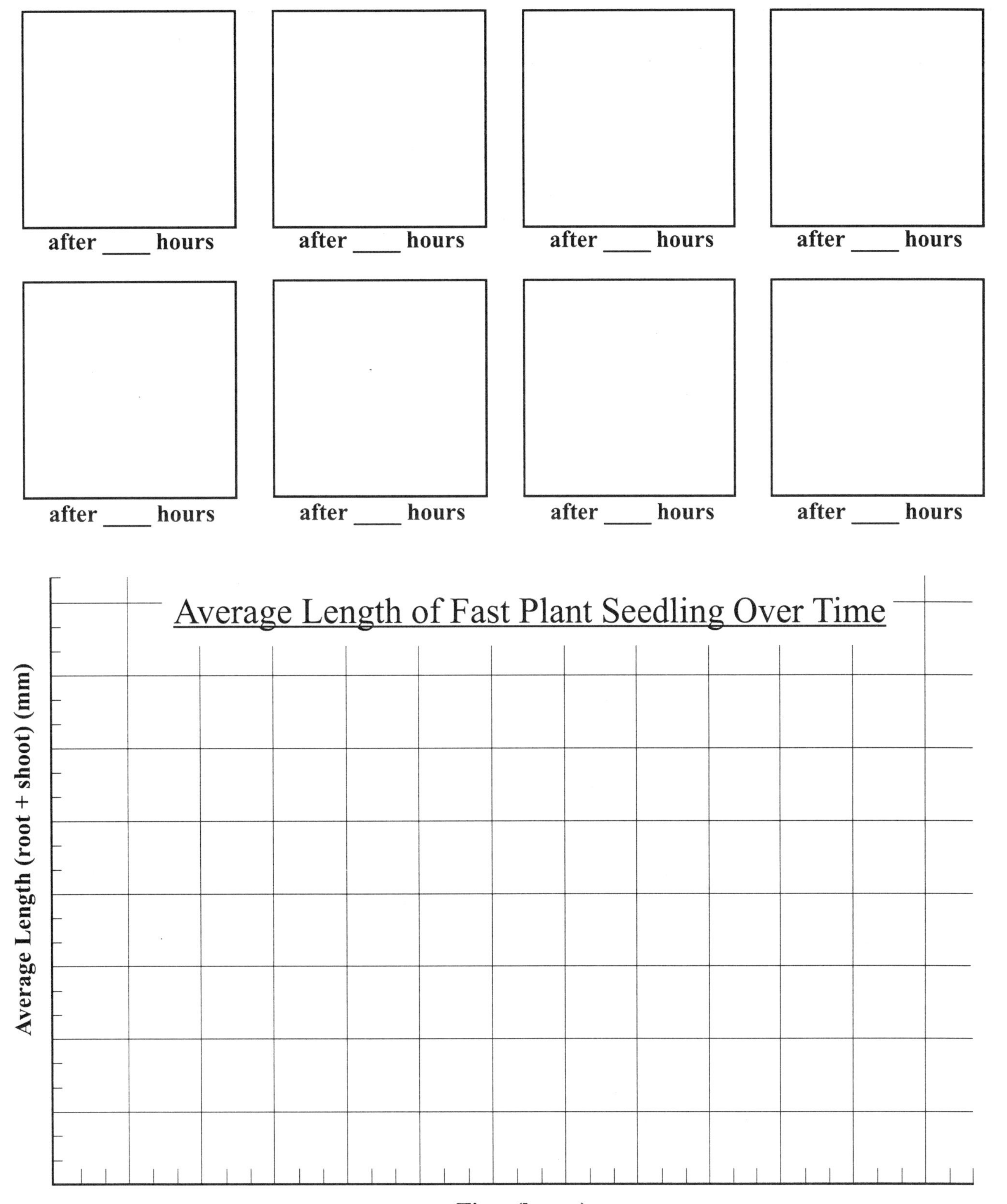

Imbibition: Swell Seeds

Individual Data Sheet

How do seeds change as they imbibe water?

Height of Seed Column

Date	Time	Hours after starting exp.	Height of seed column (mm)

Observations of seeds in different places along column: ____________________

____

____

____

____

____

____

____

____

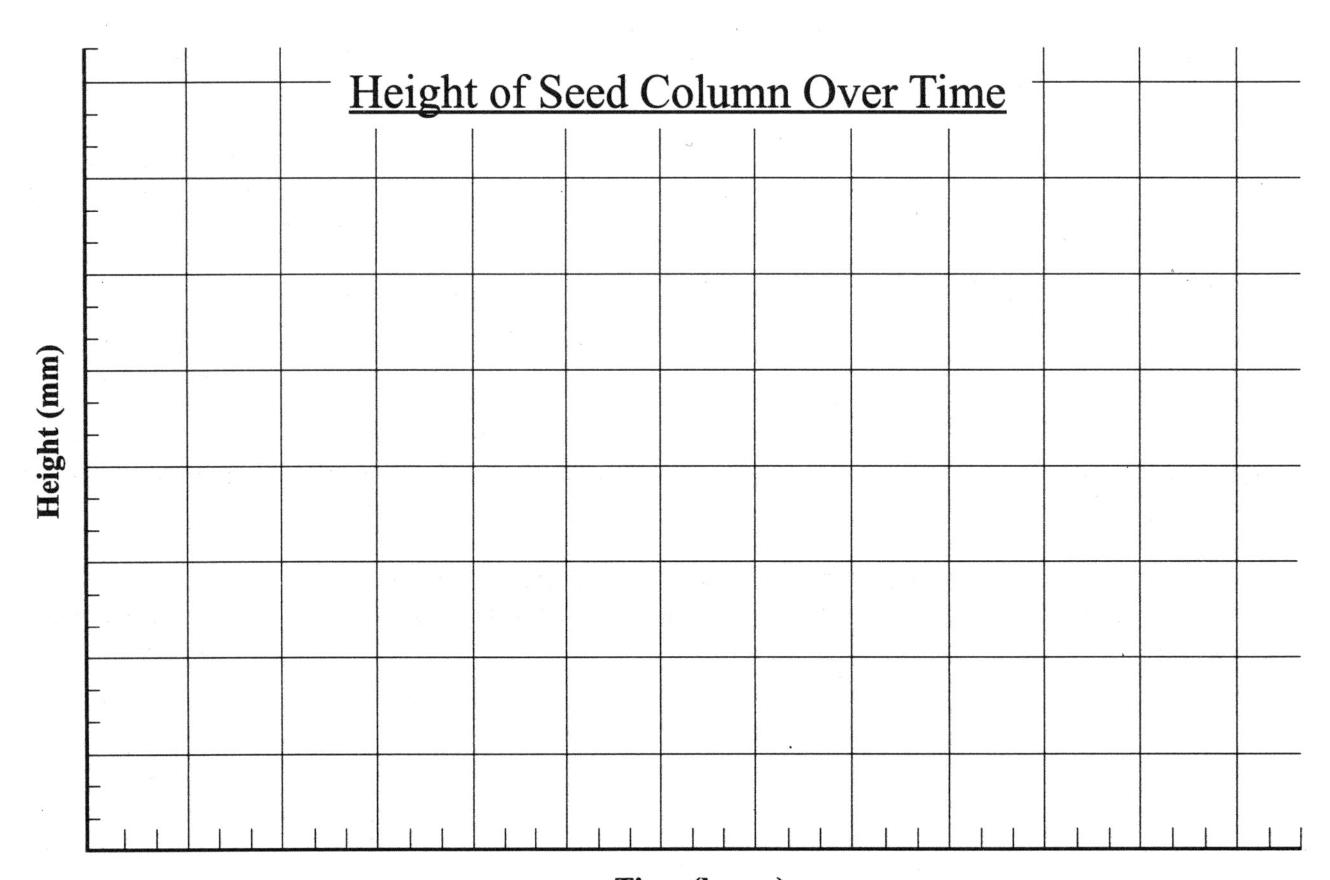

SEED DISSECTION

Sketch Sheet

Sketch of Pinto Bean and Fast Plant Seed

(indicate the magnification of your drawing)

Dry pinto bean hilum area

Soaked pinto bean hilum area

Dry Fast Plant hilum area

Soaked Fast Plant hilum area

Front view of pinto bean embryo (tip pointed down)

Germination Chamber

Sketch Sheet

Does a germinating seed "know" which way to grow?

Sketch of Seeds in Germination Chamber

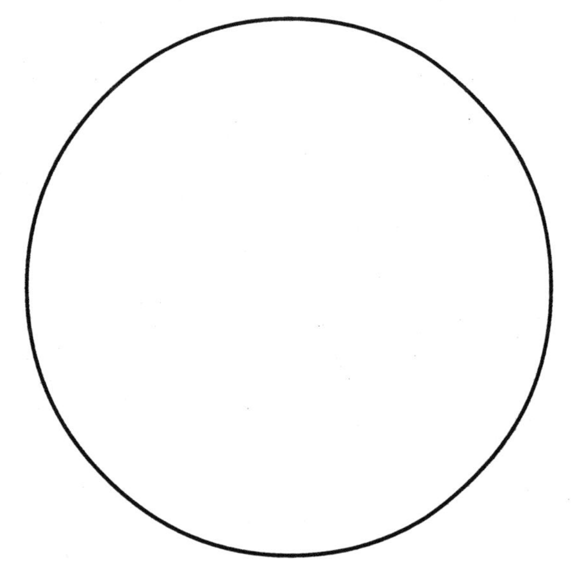

Bottle Cap Circle #1

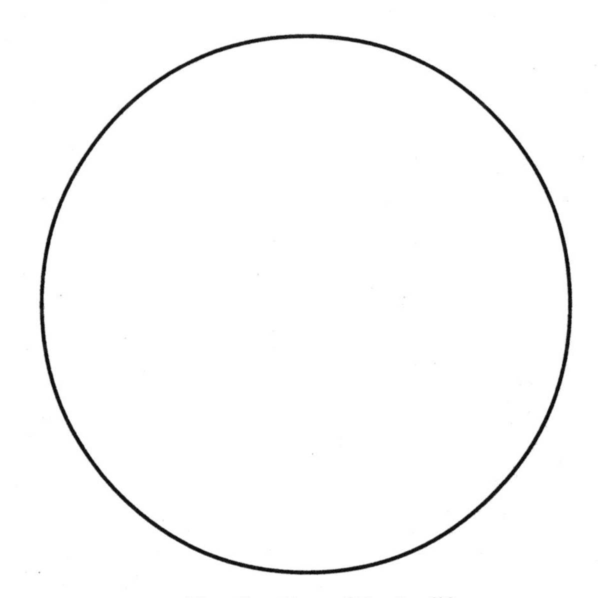

Bottle Cap Circle #2

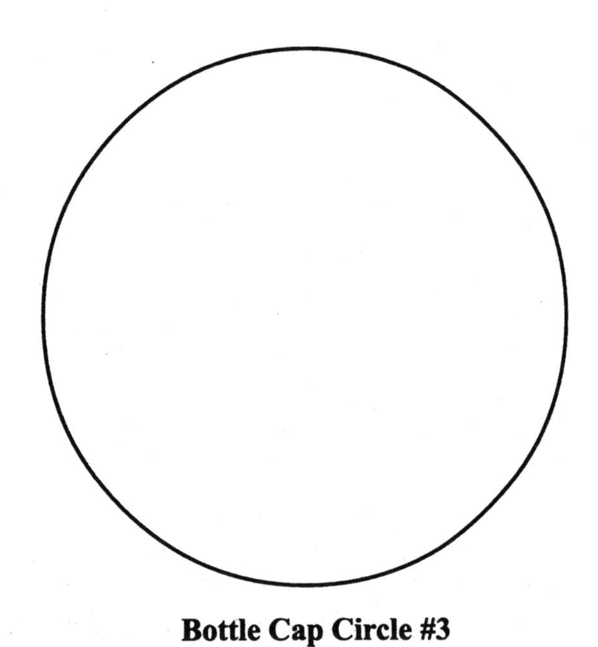

Bottle Cap Circle #3

Ruler Disk Master
for Germination Observation

copy onto
transparency sheet

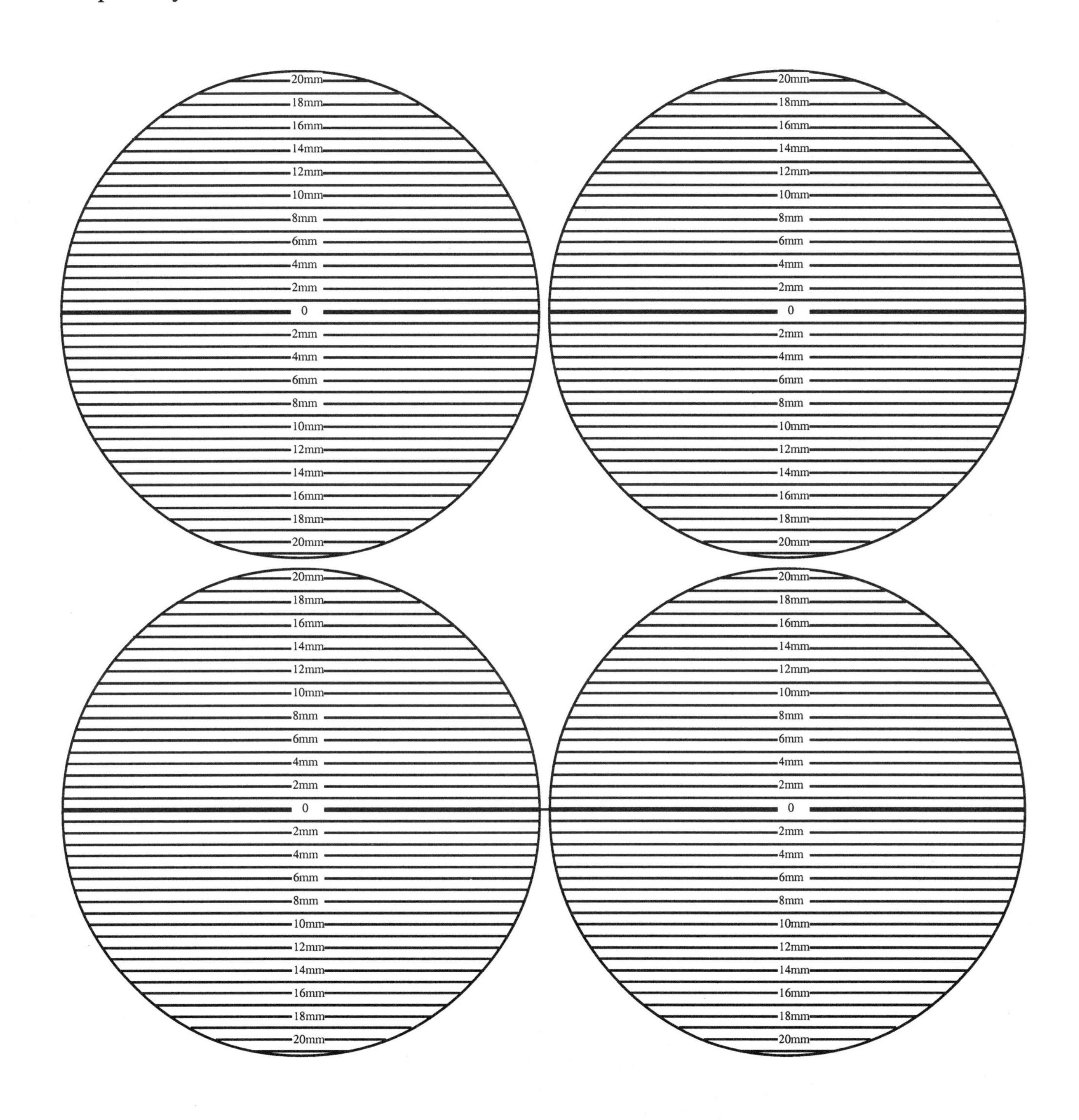

CHAPTER 2

GROWTH AND DEVELOPMENT

How much variation is there in a population of plants?

Variation among plants can be as diverse as variation among classmates!

Food for Thought...

GENES IN GENOMES

The *genome* (total sequence of DNA) of an organism is much more than the total DNA of its genes. How much more depends on the organism. The diversity of genome sizes in plants is highly variable and ranges from about 150 million bp (nucleic acid base pairs) for *Arabidopsis* to about 100 billion bp for *Trillium*. The Fast Plants genome size is about 800 million bp. The functions of the non–gene-encoding portions of a genome are only partially understood, and are an area of active research. ⊠

BACKGROUND

As a plant germinates and matures, it undergoes the processes of growth and development. *Growth* arises from the addition of new cells and the increase in their size. *Development* is the result of cells differentiating into a diversity of tissues that make up organs such as roots, shoots, leaves, and flowers. Each of these organs has specialized functions coordinated to enable the individual plant to complete its cycle in the spiral of life.

Both genetics and the environment play fundamental roles in regulating growth and development, and determining the *phenotype* (appearance) of an organism. Plants have between 20,000 and 30,000 different genes. There is small, but significant, variation among the pool of genes in each individual plant in a population. A unique set of genes defines the individual's *genotype*. Expression of these genes through growth and development causes significant observable variation in how the individual plants will appear and function during their life cycles.

GROW AND GO

Food for Thought...

While most animals reach a point in their development when they stop growing (maturity), plants usually continue to grow throughout their entire lives. In some cases, this growth can parallel the ability of an animal to move. A plant may not be able to move to a better source of light or water, but frequently it can grow toward it!

This ability of a plant to grow throughout its life is due to its *meristems*. These are undifferentiated groups of cells in a plant whose function is to give rise to new lineages of cells. These cells in turn enlarge and differentiate into tissues that serve the various specialized functions. Because meristematic tissues remain undifferentiated and can always divide, they can be considered to be "perpetually young."

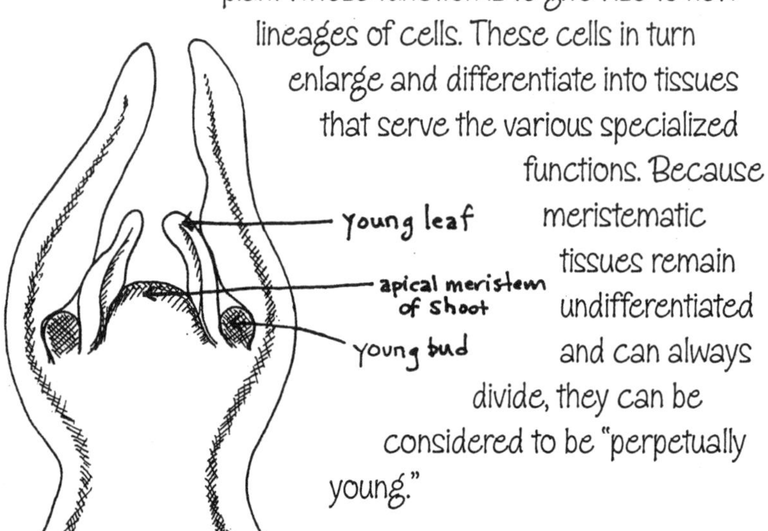

Plant growth involves a number of different meristematic tissues. *Apical meristems* are the sources of plant extension, or *primary growth*. They are found at all growing root and shoot tips. *Secondary growth* in a plant results in thickening of these organs and is the product of *lateral meristems*. Wood and bark from trees are products of secondary growth. *Axillary meristems* are found at the junction of every leaf and the stem, the outgrowth of which gives rise to branches.

INQUIRY LAUNCH

What happens when an apical meristem is removed? Does the effect vary depending on the point in the plant's life cycle?

BACKGROUND, continued

The environment is created by the interaction of physical (light, temperature, and gravity), chemical (water, air, and minerals) and biological (microbes and larger organisms) components. When overall environmental conditions are favorable, plants grow. In addition to the broad effects of the environment as a whole on a population of plants, the local conditions around individuals also lead to additional variation from one plant to another.

In Fast Plants, growth and development occur rapidly and continuously throughout the life cycle of the individual. Things are most dramatic in the 10–12 days between seedling *emergence* (arrival at the soil level) and the opening of the first flowers. Using the following activities you can explore growth, development, and variation through the Fast Plants life cycle.

ACTIVITIES

Activity 1
Tracking Variation
In this activity you will work as a team of four students with eight plants. Each team will sow seeds in wickpots and place them in an environment conducive to germination and growth. You can use any wickpot growing system that you have available (styrofoam quads, film cans, etc.) After the plants emerge you will be responsible for two of your team's plants. You will track growth and development by measuring five plant traits: plant height, number of leaves, number of hairs on the first true leaf margin, days to first open flower, and number of open flowers at specified days after sowing (*das*). All individual plant data will be recorded on the Tracking Variation team data sheet (page 56).

The data collected on each plant will become part of a class data set that you will analyze, plot and display. This will allow you to better understand the variation within a population of Fast Plants as they grow and develop.

Procedure
1. On **Day 0** (or 0 *das* or days after sowing): in teams of four, sow your team's Fast Plant seeds in each wickpot. You will want to sow twice as many seeds as you will eventually have plants. You will thin to 8 plants per team on Day 7 (see page 120).

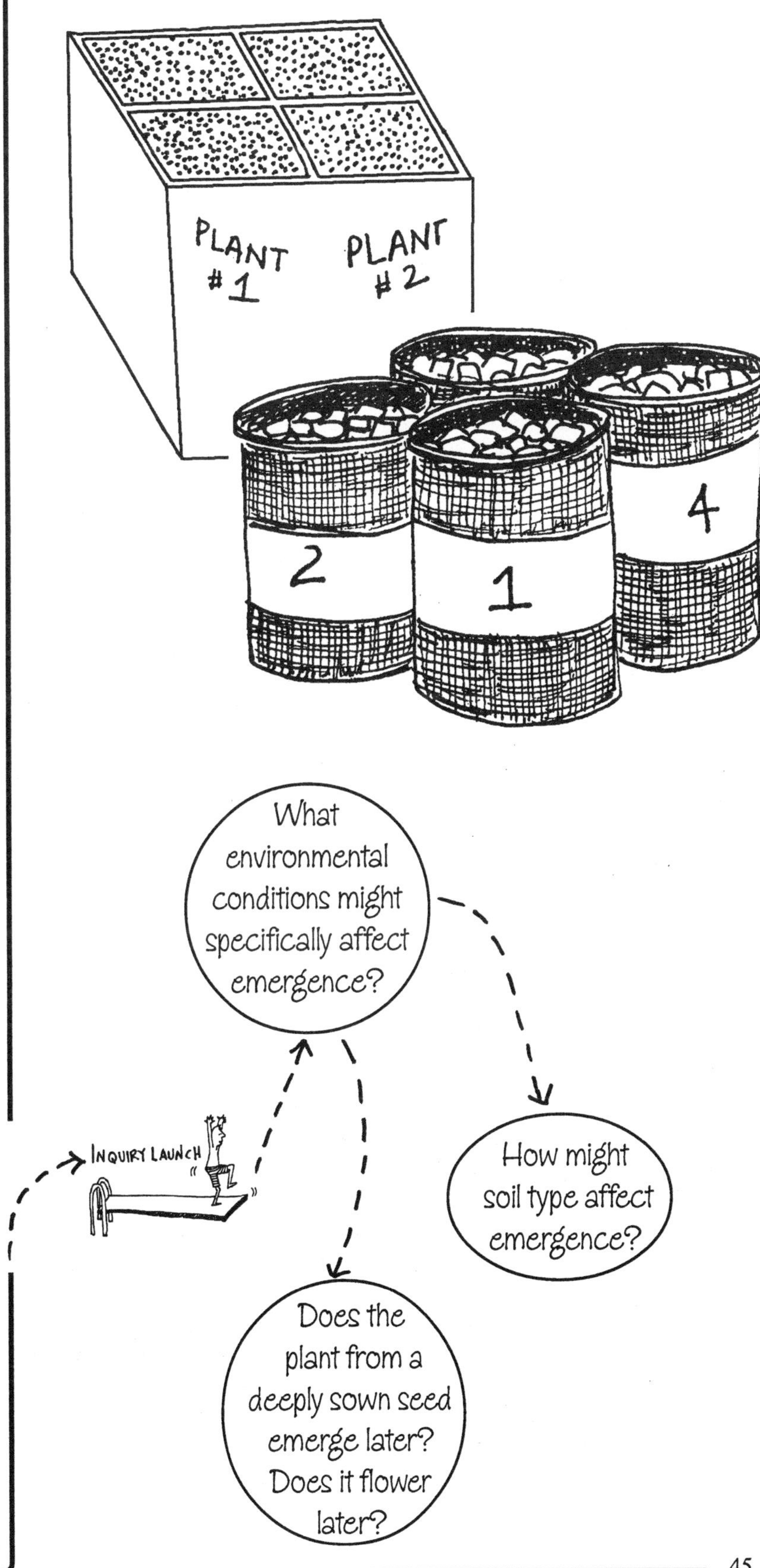

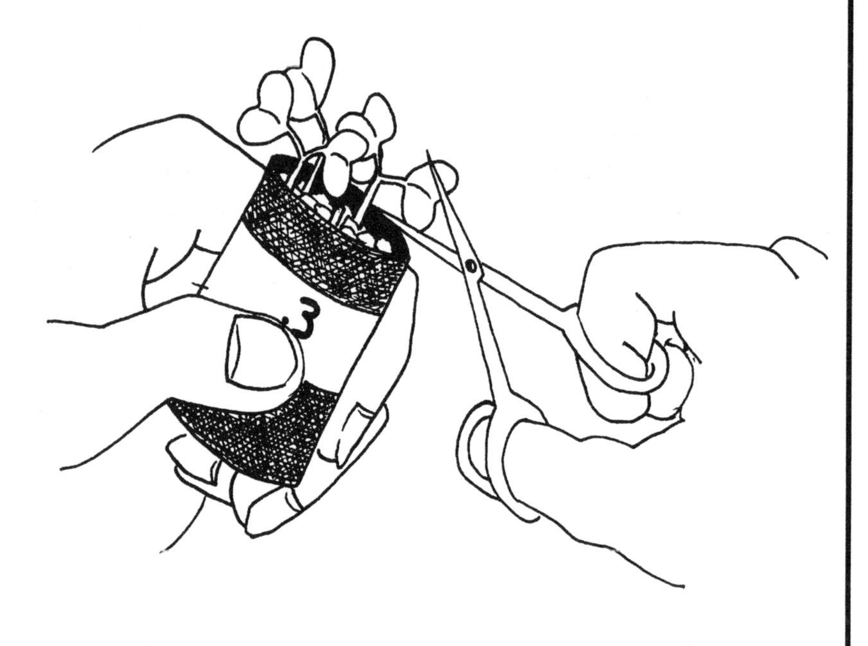

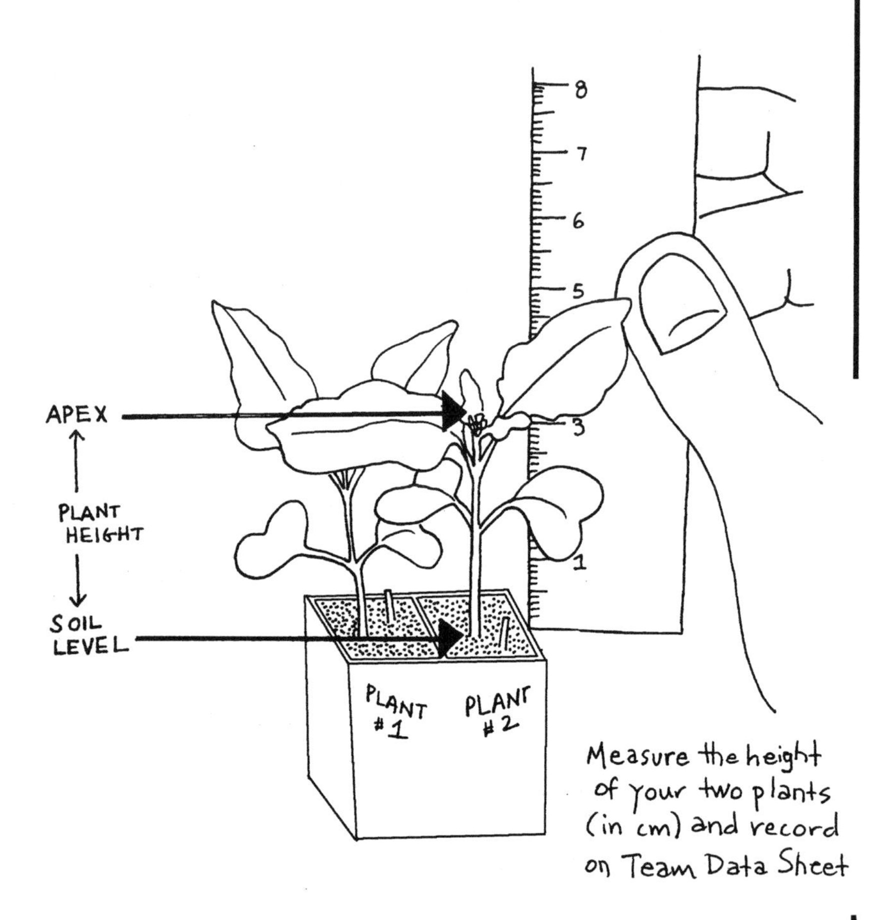

ACTIVITIES, continued

2. By **Day 3** (3 *das*), the seedlings will have emerged.
   ↪ On the Tracking Variation team data sheet (page 56) record the number of seedlings that have emerged from each wickpot. Calculate and record germination percentage numbers for yourself, the group, and the class.

3. On **Day 7**, thin to two plants per student (1 per quad cell or two per film can wickpot) by carefully snipping off the other plants at the soil level with scissors or gently pulling them out with forceps. Number the remaining plants in each wickpot. Each team should have plants numbered from one to eight.
   ↪ Measure (in centimeters) and record the height of your two plants on the team data sheet.

Notes:

- Measure height from the soil surface to the *shoot apex* (growing tip) as indicated in the illustration. Do not measure to the highest part of a leaf.
- At this time you should have your first team data set. This is a good point to practice some simple organization and analysis activities on these data and enter it onto the team data sheet.

?? Is plant height the best measure of overall plant growth? What other growth variable could/would you use?

4. Continue to observe the developing plants. As the plants grow, they will use increasingly more liquid from the system; be sure to check reservoirs daily and fill as needed.

5. On **Days 8–10**, you will notice the appearance of buds within the shoot apex of the growing tip. By **Days 8–10**, the vegetative buds have developed into floral buds. Within these floral buds the tissues that lead to the production of the male and female sex cells are developing and differentiating. Over the next six days the buds will enlarge as male gametes develop within the anthers, and as female gametes develop within the ovules of the pistil (see Chapter 3, Flowering for details on floral parts).

6. On **Day 11**, notice that the plants are growing taller. All of the leaves on the main stem have formed and the flower buds are more prominent. Count the number of leaves on the main stem including the cotyledons. Increase in leaf number is an indicator of the extent of the vegetative phase of plant development.

↪ Measure and record the height of your two plants and record the number of leaves on the main stem.

If you wish to study the flowering of the plants, the flowering activities in Chapter 3 can be carried out starting with the opening of the first flower, sometime between **Day 11** and **13**.

Notice the appearance of small floral buds in the apex of the growing shoot.

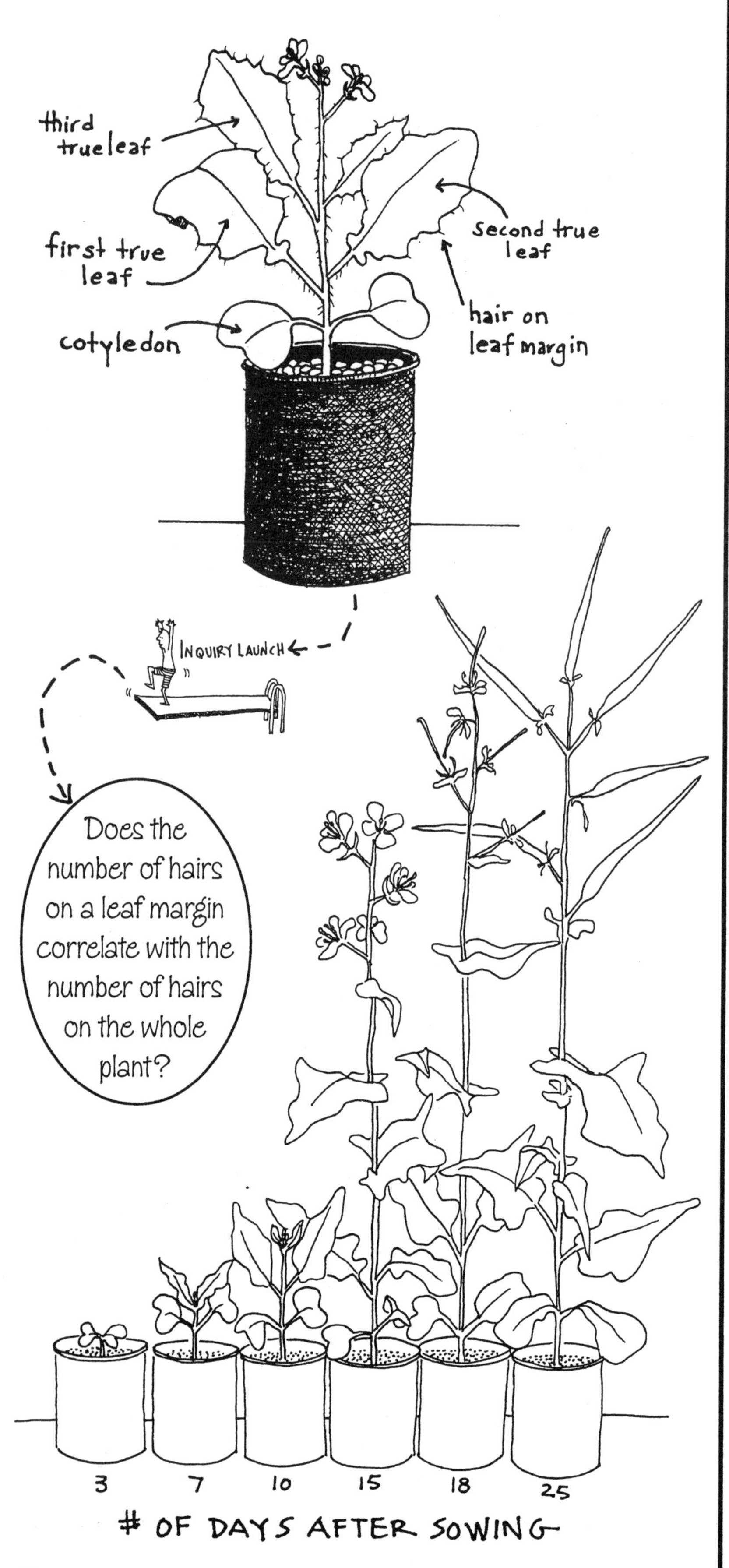

ACTIVITIES, continued

7. By **Days 12–14**, the flowers on some plants will begin to open.
   ↪ For each plant, record the number of days after sowing (*das*) when the first flower opens. (Note: The timing of flowering may vary depending on the temperature at which the plants have been growing.)

8. On **Day 14**, the plants should be flowering. With the aid of a hand lens or dissecting microscope, count the number of hairs on the *margin* (edge) of a leaf blade.
   ↪ Measure and record the height of each of your two plants and record the number of hairs on the margin of the first true leaf.

9. Following **Day 14**, many of the flowers will be opening on the plants and awaiting pollination. The pollination activities in Chapter 4 can be carried out on **Days 15–17**, or when most plants have been in flower for two days.

10. **Day 17** is generally the day of the last pollination.
    ↪ Make and record a final height measurement. Record the number of open flowers on each plant.

11. If you wish to carry out the activities in Chapter 5, Fertilization to Seeds, you should terminate flowering as described on page 96 in Chapter 3, Flowering.

12. Combine the team data into a class data summary using the class data sheet. All height data should be recorded on the Plant Height class data sheet (page 57). Other data should be recorded on the Additional Traits class data sheet

(page 58). Notice how the various statistical factors (range, mean, and standard deviation) change over time from sowing. (See Appendix 2, page 127, for a discussion of statistical measures.)

13. Use data analysis calculations (and software if available) to create graphical and statistical summaries of the class data. Consider the following questions:

?? How does variation in the plant height data change as the plant develops?

?? Which of the traits you measured (height, number of seedlings emerged by 4 *das*, number of leaves per stem, days to first flower, hairs on first leaf margin, or number of open flowers by 18 *das)* were the most variable?

?? Are the means from your group different from the class means? Which do you think are more accurate and why?

?? What could be causing the variation in your class population of plants?

14. The Tracking Variation: Frequency Distribution class data sheet (page 59) provides a master graph for a set of class histograms. On a copy of the graph, create a class frequency histogram for plant height at 7 *das*. Create three more class histograms for height at 11 *das*, 14 *das*, and 18 *das*. (For more information on histograms see Appendix 2, page 127.)

?? From a class frequency histogram and statistical summary, does the measured plant character of height exhibit a normal distribution within the class population?

THOUGHTS ON STATISTICS

Food for Thought...

The quiet statisticians have changed our world—not by discovering new facts or technical developments but by changing the ways we reason, experiment, and form our opinions about it.

—I. Hacking

All knowledge is, in the final analysis, history.
All sciences are, in the abstract, mathematics.
All judgments are, in their rationale, statistics.

—C. R. Rao,
Statistics and Truth

ACTIVITIES, continued

?? Do your individual plant heights fall within one standard deviation of the class mean? Would you consider your plant heights to be normal? Why or why not?

If desired, class histograms can be created for the other plant traits measured.

?? Judging from the class histograms created for other measured traits, compare plant variation with respect to different traits.

?? Are there any "abnormal" plants?

Activity 2

The Hypocotyl Hypothesis

Gravity is a key environmental factor in plant growth. However, the way it affects different plant tissues can vary significantly. At some developmental stages, certain tissues are very responsive to changes in gravity. This activity explores the effect of gravity on hypocotyls. The *hypocotyl* is the embryonic stem of the seedling that extends between the *radicle* (embryonic root) and the *cotyledons* (seed leaves).

During germination and emergence of the seedling, the hypocotyl grows rapidly by extending the cotyledons, and is very responsive to temperature, light, and gravity. This activity takes advantage of hypocotyl sensitivity to explore *gravitropism* (plant growth movements in response to gravity) with Fast Plants.

Procedure

1. Place a paper towel wick strip on the inside of a black film can (Figure 1).

2. Wet the paper towel. Leave no more than three to four drops of water at the bottom of the can (Figure 2).

3. Cut a three- to four-day-old seedling at soil level. This should leave the majority of the hypocotyl attached to the cotyledons.

4. Pull the wick strip 1/3 of the way out of the can and place the cotyledons of your cut seedling on the strip, such that the hypocotyl is facing away from the strip (Figure 3).

5. Push the strip back into the can until it is below the opening lip of the can. (A pencil tip is a good pushing tool for this.) Snap on an opaque lid (Figure 4).

6. Develop a *hypothesis* (predicted result and explanation) for how you would expect the hypocotyl to respond after 4–24 hours. Write down your hypothesis and reasoning.

7. Record your results with a drawing after 4–24 hours. (Peeking is permitted!)

8. Compare, pool, and analyze your results with the rest of the class.

?? Why did the plants respond as you observed? Design an experiment to test your hypothesis.

?? What environmental factors could have influenced your results? How can you test your hypothesis?

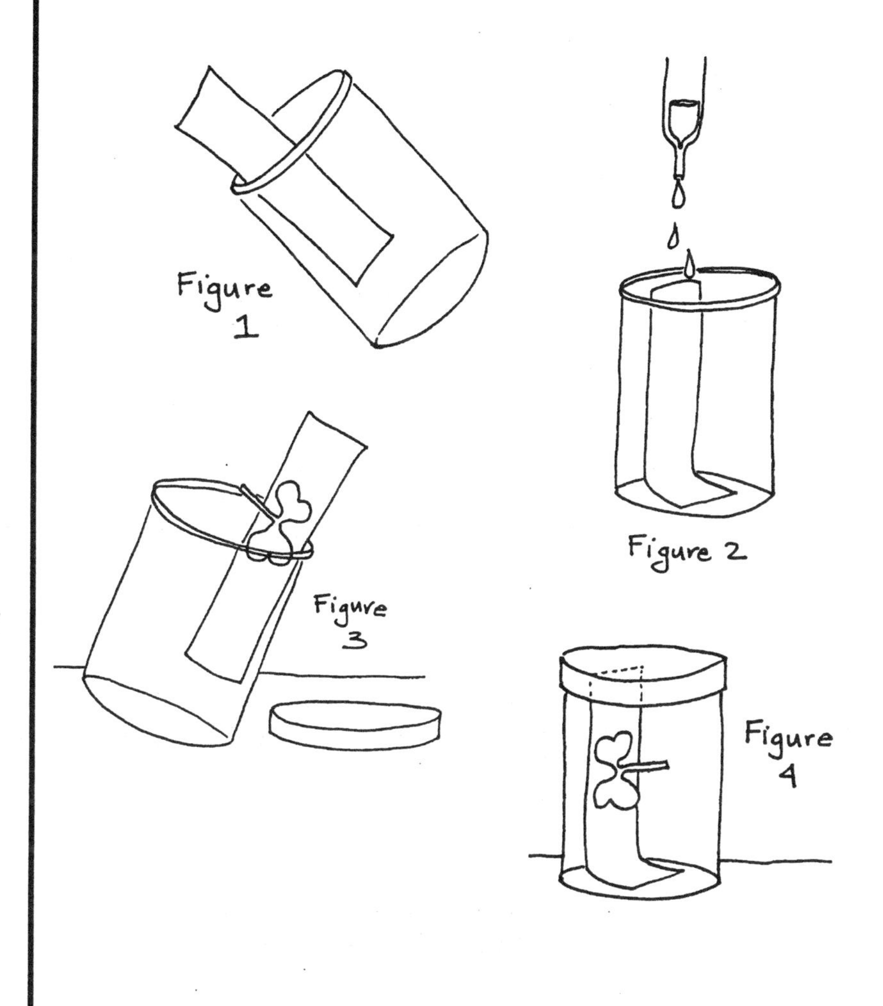

Classroom Vignette: Mr. B., 7th Grade

A group of three students approaches Mr. B. with their film cans in hand. "Mr. B., we have the seedlings stuck to the sides of the film can by their hypocotyls..." Scott glances at the drawing on the board, "—their cotyledons. Do we need to cover the open end with a lid?"

Avoiding giving an answer, Mr. B. coaxes the students to think further. "Well, what do you think?"

"If we don't cover it, it could dry out?"

"That's a thought....Katie, what do you think?"

"Wouldn't we also want a cover on the film can to eliminate light?"

"Oh yeah!" The light bulb goes off in Scott's head. "We want to remove light as a variable!"

He turns to the class and addresses them as a whole. "Remember! When you are done, you need to make a prediction. Copy the frontal view into your lab notebooks. Then draw what you expect the hypocotyls to look like by tomorrow. Stick your necks out! Make a prediction!"

Bobby interjects, "You mean like all limp and dead?" His friends snicker.

"If that's what you think, then write that down. But it won't be an easy way out because you must back up your prediction with some solid reasoning."

"No! I'm not slackin'. Seriously! We killed them! They are dead, so—"

Mr. B. cuts in, "Is that an observation, or an inference?"

A sprinkling of kids chimes in, "Inference!!!"

"OK! OK! Hold on to those thoughts. Journal on it if it helps you sleep tonight. So, maybe we might want to clarify. 'Assuming they are still alive, I'd expect...' Write a brief passage telling why you think they will do this! You cannot open your canister unless you've made a prediction. No peeking and no talking to other classes!"

See you in the next chapter for FLOWERING!

See Appendix 5 (page 145) for the full vignette.

Additional Reading

J. E. Dale, "How Do Leaves Grow?" *BioScience* **42**:425–431, 1992. Concentrates on the development of individual leaves and their function, and genetic and environmental control mechanisms of this development.

C. Darwin & F. Darwin, *The Power of Movement in Plants*, Murray, London, 1880.

R. D. Firn & J. Digby, "The Establishment of Tropic Curvatures in Plants," *Annual Review of Plant Physiology* **31**:131–148, 1980. Reviews shoot geotropism and phototropism, and root geotropism.

R. A. Kerstetter & S. Hake, "Shoot Meristem Formation in Vegetative Development," *The Plant Cell* **9**:1001–1010, 1997.

I. R. Macdonald, et al., "Analysis of Growth During Geotropic Curvature in Seedling Hypocotyls," *Plant Cell and Environment* **6**:401–406, 1983. Demonstrates the inhibition of growth along the upper side of a hypocotyl undergoing geotropic curvature and the simulation of growth on the corresponding underside.

R. S. Poethig, "Phase Change and the Regulation of Shoot Morphogenesis in Plants," *Science* **250**:923–930, 1990.

C. R. Rao, *Statistics and Truth: Putting Chance to Work*, International Publishing House, Fairland, Maryland, 1989.

W. K. Silk, "Quantitative Descriptions of Development," *Annual Review of Plant Physiology* **35**:479–518, 1984. Focuses on the quantification of plant growth.

Overview

The two activities in this chapter provide you and your students with opportunities to explore the concepts of plant growth, development, and variation with Fast Plants. In Activity 1, students follow the maturation of Fast Plants from emergence of the seedling to flowering. During this time period students will track height and record the date of emergence, number of leaves on the stem, the opening of the first flower, the number of hairs on the margin of the first true leaf, and the number of open flowers by 18 *das*. Critical to this activity is analysis and reflection of individual and composite data as the students consider the question, "How much variation is exhibited within and among subpopulations of Fast Plants grown under normal conditions?" Students consider which traits are most variable as well as which life cycle stages display the greatest variation for a single, given trait. In Activity 2, students explore the effects of gravity on Fast Plant hypocotyls.

Objectives

In participating in this activity students will

- observe, measure, and record plant growth and key developmental events from seedling emergence through flowering;
- understand the role of environment in regulating plant growth;
- observe, measure, and analyze variation in growth and development among individuals in a population of plants;
- consider the use of statistical and graphical representation of growth and development within a population; and
- understand that growth in plants represents an ordered sequence of developmental events, which vary between individuals of a population within limits that are defined as "normal" (see Appendix 2, page 127).

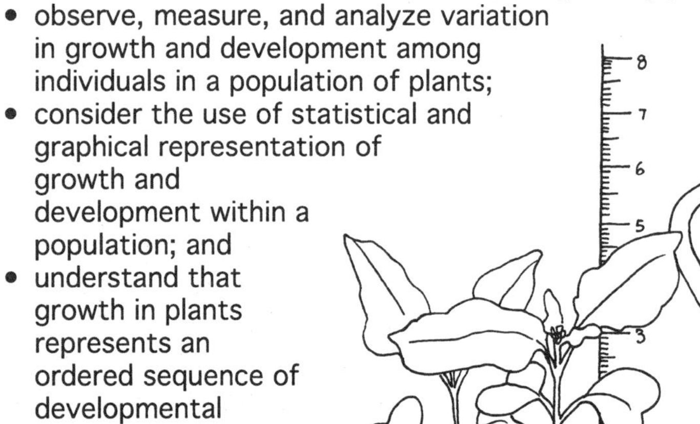

Time Required

Activity 1:
Tracking Variation

- **Day 0** (0 *das*): one 50-minute class period to plant seeds.
- **Day 3** (3 *das*): 5 minutes to record seedling emergence.
- **Day 7**: 20 minutes to thin seedlings, label and measure remaining plants, and record data.
- **Day 11**: 10–15 minutes to measure plants, count leaves, and record data.
- **Days 12–14**: 5 minutes to check plants and record opening of first flower.
- **Day 14**: 30 minutes to measure plants, count hairs, and record data.
- **Day 17**: 10 minutes to measure plants, count flowers.

The above reflects only the time necessary for mechanics and recording data. Additional time is necessary for data analysis and discussion.

Activity 2:
The Hypocotyl Hypothesis

- **Day 1**: 30 minutes to set up experiment and record hypothesis.
- **Day 2**: 15 minutes to examine, tabulate, and discuss results.

Materials

Activity 1

Each group of 4 students will need:

- white lab tape
- metric ruler
- black fine-tipped marking pen
- fine scissors
- rubber band
- 10X hand lens or dissecting microscope
- reservoir, wicks, and nutrient solution or water*
- wickpots, wicks, and planting medium*
- Fast Plant seeds
- lighting system*

Activity 2

Each student will need:

- black film can with opaque lid
- eye dropper or small pipet
- absorbent paper toweling wick (1.0 cm x 4.5 cm)
- 1 three day-old Fast Plant seedling*

* See Appendix 1 (page 113) for growing instructions.

Classroom Management Tips

- Consider pollinating plants and continuing to record growth and development of the plant post-fertilization.
- In the Hypocotyl Hypothesis activity the water on the wick should hold the seedling in place. If the seedling is reluctant to stick, add an additional drop of water on the wick. When snapping on the lid, make sure the ends of the wicks don't protrude out of the can. Handle the chambers gently.
- To make wick strips, fold the paper towel and cut several layers at once.

Fast Plants: Growth and Development Activities

MONDAY	TUESDAY	WEDNESDAY	THURSDAY	FRIDAY
				PLANT 0 das
Record # of seedlings emerged 3 das	4 das	5 das	6 das	→ thin plants → measure height 7 das
10 das	→ measure height → count leaves 11 das	Plants begin to flower — Note das for first flower 12 das	13 das	→ measure height → count hairs on leaf margin pollinate if desired 14 das
pollinate if desired 17 das	→ measure height → count open flowers → terminate flowering 18 das			

TRACKING VARIATION
Team Data Sheet

Student Name 1 ____________________
Student Name 2 ____________________
Student Name 3 ____________________
Student Name 4 ____________________
Group Number ______

Environment
Distance in cm of plants from bulbs: ________
Average daily temperature of growing environment: _____°C
Nutrient used: ____________________
Seed type: _________

das	Character/Activity	Plant Measurements								Statistics			
	Students:	**Student 1**		**Student 2**		**Student 3**		**Student 4**		**Team germ %**		**Class germ %**	
3	seedlings emerged & germ %												
	Plant Number:	**1**	**2**	**3**	**4**	**5**	**6**	**7**	**8**	**n**	$\bar{x}$	**r**	**s**
7	plant height (cm)												
11	plant height (cm)												
11	number of leaves on stem												
14	plant height (cm)												
14	# of hairs on leaf margin												
17	plant height (cm)												
17	number of open flowers												
	day to first open flower (*das*)												

das = days after sowing, n = number of measurements, $\bar{x}$ = mean (average), r = range (maximum minus minimum), s = standard deviation

How does variation in the plant height data change as the plant develops?

TRACKING VARIATION

Class Data Sheet

Plant Height

How is the change in range and standard deviation from 7 to 17 *das* reflected in the corresponding histograms?

Date ______________________
Teacher Name ______________________
Class Period ______________________

Environment
Distance in cm of plants from bulbs: ________
Average daily temperature of growing environment: _____ °C
Nutrient used: ______________________
Seed type: _________

	7 *das*				**11 *das***				**14 *das***				**17 *das***			
Team Statistics:	n	$\bar{x}$	r	s	n	$\bar{x}$	r	s	n	$\bar{x}$	r	s	n	$\bar{x}$	r	s
Group 1																
Group 2																
Group 3																
Group 4																
Group 5																
Group 6																
Group 7																
Group 8																
Group 9																
Group 10																

das = days after sowing, n = number of measurements, $\bar{x}$ = mean (average), r = range (maximum minus minimum), s = standard deviation

TRACKING VARIATION

Class Data Sheet

Additional Plant Traits

Date ____________________

Teacher Name ____________________

Class Period ____________________

Environment

Distance in cm of plants from bulbs: ________

Average daily temperature of growing environment: _____ °C

Nutrient used: ____________________

Seed type: __________

	# of leaves on stem at 11 *das*				**# hairs, leaf margin at 14 *das***				**# open flowers at 17 *das***				**days to first flower (*das*)**			
Team Statistics:	**n**	**$\bar{x}$**	**r**	**s**	**n**	**$\bar{x}$**	**r**	**s**	**n**	**$\bar{x}$**	**r**	**s**	**n**	**$\bar{x}$**	**r**	**s**
Group 1																
Group 2																
Group 3																
Group 4																
Group 5																
Group 6																
Group 7																
Group 8																
Group 9																
Group 10																

das = days after sowing, n = number of measurements, $\bar{x}$ = mean (average), r = range (maximum minus minimum), s = standard deviation

Tracking Variation

Class Data Sheet

Frequency Distribution

Measured trait ______________________

Age of plant ______________

$n =$ ______________

$\overline{x} =$ ______________

$r =$ ______________

$s =$ ______________

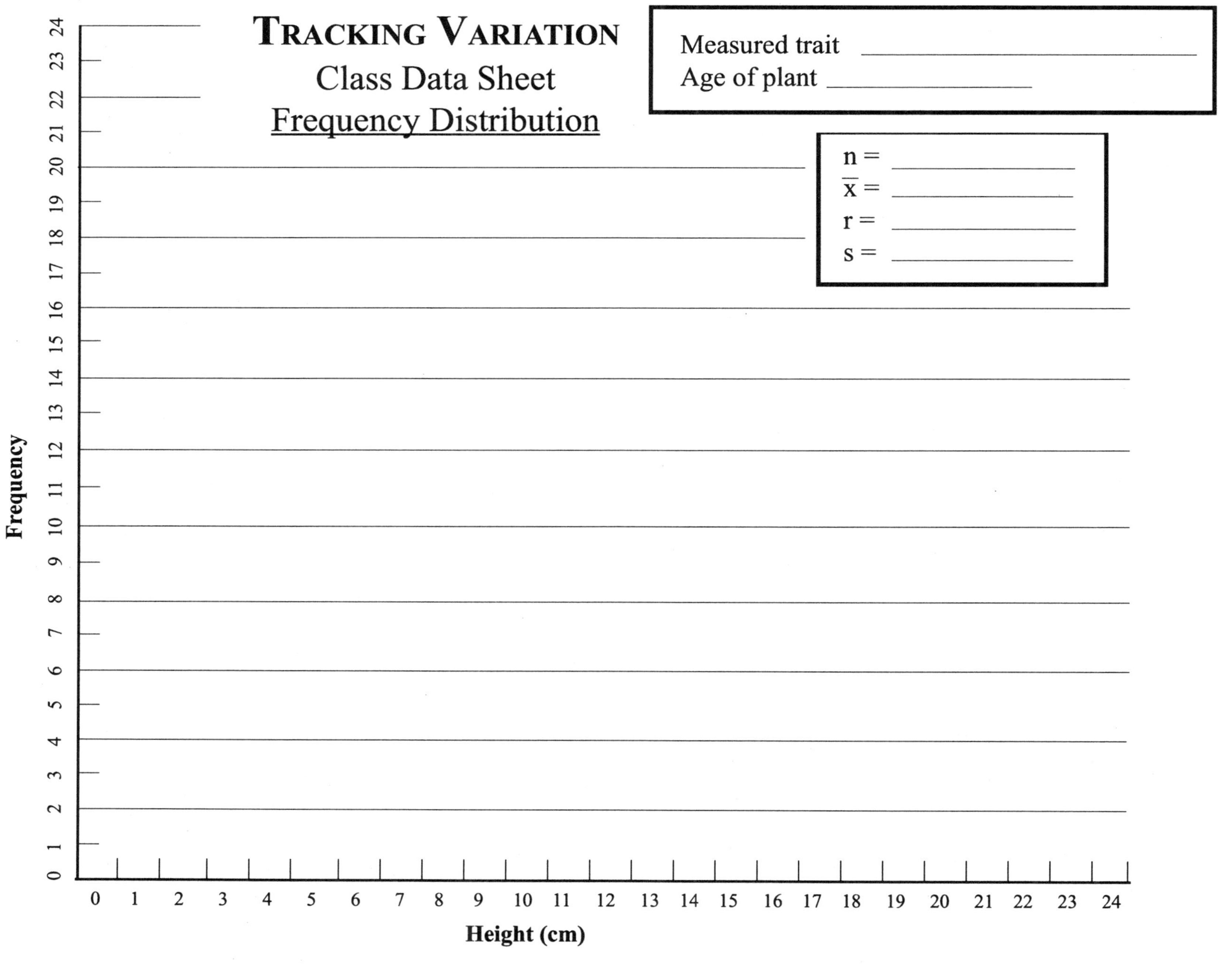

$\overline{x}$ = mean (average), r = range (maximum minus minimum), s = standard deviation

CHAPTER 3

FLOWERING

How do plants get ready to reproduce?

GROWTH CONTROL

The production of flowers by a plant represents a major shift in its development. Flowers look unlike any other part that a growing plant has produced since germination. Still, flowers share their origins with other plant parts such as leaves.

As discussed in Chapter 2, Growth and Development (page 43), all of the aboveground parts of a plant originate from shoot meristems. In the past several years, researchers have begun to unravel the mechanisms that signal the shoot meristem to shift from vegetative to floral development. Much of this work has utilized a bizarre collection of mutant plants that have altered floral parts. Some mutants are missing petals or stamens, while others have leaves in place of these parts.

By studying these mutants, a complex set of genes and corresponding proteins have been identified that control the development of flowers. These mutants have also demonstrated that floral organs are really intricately modified leaves.

BACKGROUND

What is a Flower?

What is a flower? In human eyes, it is something to enjoy, with colorful petals and fragrance. However, for many plants, the critical part of the flower is not the dramatic blossom. Within that blossom are the organs of reproduction that allow the plant to reproduce sexually and create offspring slightly different from itself.

Many plants can arise directly from an existing plant through asexual reproduction. However, when a leaf cutting sprouts new roots or an iris plant is divided, this asexual reproduction gives rise to offspring that are genetically identical to the parent plant. Hence, asexual reproduction will not generate the variation necessary to allow the species to slowly adapt to the environmental changes that will inevitably occur.

Sexual reproduction requires the union of two gametes, a male sperm and female egg, to form a *zygote* (fertilized egg). Uniting eggs and sperm from different flowers or

NIGHT LIFE

Food for Thought...

Like many cultivated varieties of plants, Fast Plants will flower regardless of time of year. But for many plants, flowering too early or too late in the year can spell disaster for the plant or its offspring. How does a plant know when to flower? One of the main ways that plants sense the right seasonal moment is by day length or, more accurately, by sensing the length of the night. An extension of this is that a short burst of light in the middle of the night will cause some plants to respond as if they had been exposed to a long day and short night!

BACKGROUND, continued

different plants provides a challenge. Plants, which are largely immobile, have evolved strategies to move their male gametes long distances to fertilize the female gametes. One common strategy involves employing animals, often insects, to carry *pollen* (male gametes) to the *pistil* (female reproductive organ).

In order to attract the insects into such service, the plants provide food, in the form of nectar or pollen. However, the plant must first attract the insects. This attraction must happen when the reproductive organs within a flower are ready to provide and receive pollen. Plants have evolved a constellation of intriguing features by which they can "advertise" the availability of pollen and nectar to the pollinators. These "advertisements" include familiar flower characteristics such as dramatic colors and color patterns, distinctive fragrances, and large or complex shapes. The flower advertises the availability of nectar, which lures the pollinators into service.

So the answer to the question *"What is a flower?"* is a matter of the perspective of the viewer. For an insect, the flower is an essential source of sugar-rich nectar and protein-rich pollen. To humans, it is a delightful gift of beauty. But for plants, the flower is the means by which they are able to generate, through sexual reproduction, the variation necessary for evolution and survival of their species.

Brassica FLOWER PARTS

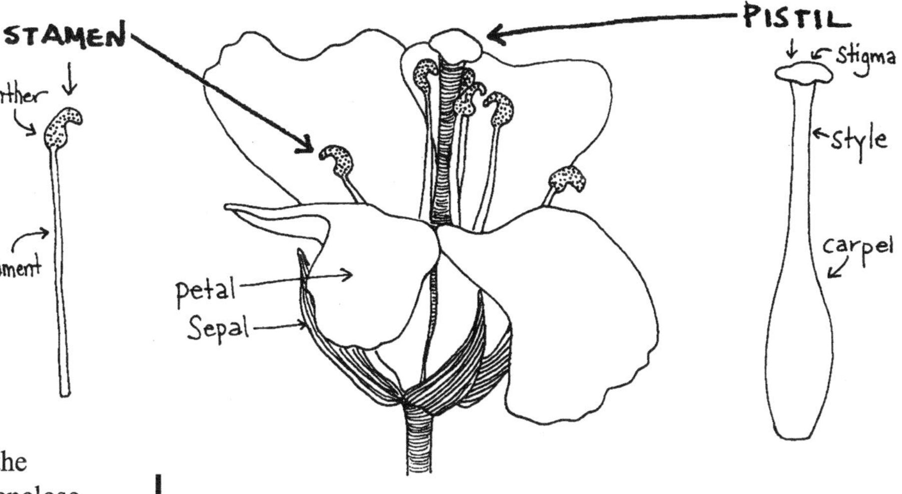

Inside the Flower
Most flowers have the same basic parts, though they are often arranged in different ways. Each of the four main parts of a flower, the *sepals*, *petals*, *stamens*, and *pistil* serve particular functions in flowering and sexual reproduction. The sepals are the green leaflike structures that enclose and protect the developing flower. The petals are the colored leaflike structures that lie within the ring of sepals and frequently serve to attract pollinators.

The stamen consists of the *filament*, a slender stalk upon which is borne the *anther.* Within the anther are the pollen grains, which contain the male gametes or sperm cells.

The pistil usually has three parts, the *stigma* (which traps the pollen), the *carpel* (ovary), and the *style* (the neck between the two). Brassica (Fast Plants) flowers have two carpels fused together and separated by a thin membrane. The carpels house the *ovules*, each of which contains the female gametes.

In brassicas and many other species that need to attract specific pollinators, *nectaries* are also present. These nectaries, strategically located in the flower, secrete sugar-rich nectar. Their location ensures that nectar-gathering insects and other animals will receive pollen from anthers and transmit it to its stigmas as they forage.

ALICE NEVER HAD IT SO GOOD...

Food for Thought...

"It really doesn't matter whether you can draw or not—just the time taken to examine in detail, to turn a flower or a shell over in your fingers, opens doors and windows. The time spent observing pays, and you can better observe with a hand lens than without one. A hand lens is a joy and a delight, an entree to another world just below your normal vision. Alice in Wonderland never had it so good—no mysterious potions are needed, just a ten- or fifteen-power hand lens hung around your neck. There's a kind if magic in seeing stellate hairs on a mustard stem, in seeing the . . . barbed margin of a nettle spine—there all the time, but never visible without enhancement. But to take the next step—to draw these in the margin of your notebook, on the back of an envelope, in a sketch pad, or even in the sand—establishes a connection between hand and eye which reinforces the connection between eye and memory. Drawing fastens the plant in memory." ⊠

— Ann Zwinger from Finding Home

IT'S ALL RELATIVE (HUMIDITY)

Food for Thought...

If the atmospheric relative humidity is very high (>95%), mature anthers in flowering Fast Plants may fail to open (*dehisce*) to expose their pollen. This occurs when plants are grown in closed containers in which relative humidity builds up. It can be remedied by circulating air over the plants with a fan; mature anthers will then usually dehisce within a few minutes.

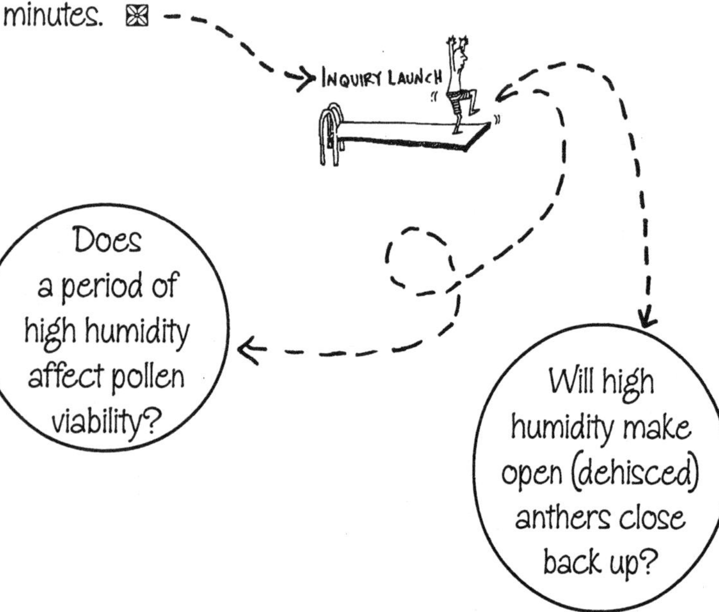

BACKGROUND, continued

Fast Plants and Flowering

Several days before the flowers open, deep in the apical bud whorl, meiosis is occurring in the anthers and ovules. Meiosis is responsible for the variation that results from sexual reproduction.

About 7 days after sowing (*das*) your Fast Plants, you will notice a tightly packed whorl of buds, some of which are larger than others. Each successively smaller bud will open after its predecessor, resulting in a beautiful sequence of flowering. Normally the lowest flower on the shoot opens, followed by the next highest, and so forth. The first flower will open 13–15 days after sowing.

When is a Flower Open?

Determining exactly when a flower is open is not as straightforward as it might appear. As you observe the progression of flowers opening, you will notice that, as the buds swell, the sepals are pushed apart by the enlarging anthers and emerging yellow petals. Eventually the petals (which are rolled) fold outward about halfway up their length, flattening and spreading to reveal their bright yellow color. At this time you might conclude that the flowers are fully open.

A flower is fully functioning when the petals are completely opened, the reproductive organs are ready to produce and receive pollen, and nectar is being produced. Taking the plants' perspective, a flower is defined as fully open when it is capable of providing and receiving pollen. Thus, not until the anthers *dehisce* (open up; Latin for *to open*) and release their pollen is a flower functionally open.

The shedding of pollen is known as *anthesis* (Greek for *full bloom*). When you observe a succession of flowers on your Fast Plants you will observe whether anthesis has occurred by noting the release of the powdery yellow pollen from the anthers. Consider a flower to be fully open when anthesis first occurs. A hand lens can be helpful in detecting anthesis.

ACTIVITIES

Activity 1
The First Flowers

1. Sometime between **Day 13** and **Day 15**, flowers on individual plants will begin to open. Note the number of days after sowing (*das*) when the first flower opens. Consider a flower to be open when anthesis first occurs.

↪ Record the opening of the flowers on the data sheet The First Flowers (page 72).

?? How much variation exists in time to first flower within the population of Fast Plants in your classroom?

?? What is the timing and sequence of the opening of the next flowers on the shoot?

2. Starting with the oldest, lowermost flower on the main stalk, number the flowers as 1, 2, 3, 4, and 5 with number 1 being the oldest flower. With a sharp pair of fine scissors, terminate flowering by snipping off all remaining flowers above flower number 5. Snip off all side shoots. On subsequent days, you may need to terminate further flowering by snipping new buds and shoot apices, and side shoots.

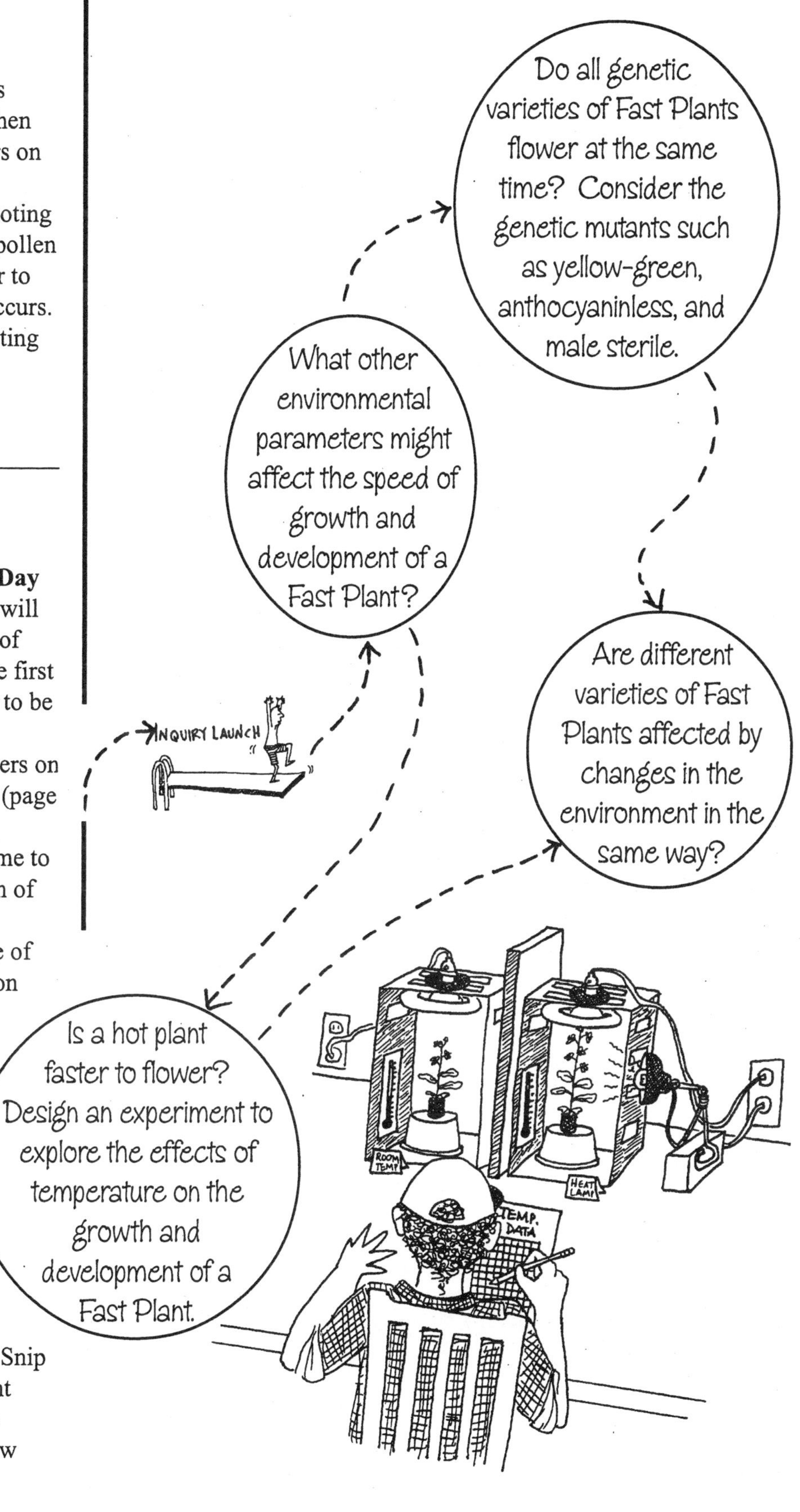

Activities, continued

Activity 2
The Flower Spiral
After terminating flowers in Activity 1, note that the flowers open in a sequence spiraling up the plant stem. On the Floral Spiral individual data sheet (page 73), draw the location of the flowers on the stem. Measure the angle between successive buds and enter your data on the individual and class Flower Spiral data sheets (pages 73 and 74).

?? What is the pattern of flowers on the stem? What is the angular distance between two successive flowers? Is the angle constant for all flowers on a plant?

Activity 3
What's in a Flower?
Take the *top open flower* of the first plant and carefully remove it with a forceps.

While observing with a hand lens or microscope, carefully remove the flower parts with fine tipped forceps, a dissecting needle, or a sharp-tipped toothpick. For each part, note the relative positions on the receptacle. Place each part on the sticky side of a piece of tape, taped to a card as shown in the illustration. Refer to the illustration on page 63 for help in identifying the floral parts.

?? Compare the Fast Plant's flower parts to those of another flower such as a tulip or lily.

FLOWER DISSECTION

BRASSICA FLOWER PARTS
petals sepals stamens pistil

Carefully remove the flower parts, placing each piece on sticky tape mounted on an index card.

Activity 4
Orientation of Floral Parts
The relative arrangement of the basic floral parts is important in understanding how a flower functions, as well as in taxonomic identification of the plant. The relative location of floral parts can be illustrated with a floral diagram.

A floral diagram represents a **cross section** of all the parts of a flower as it would appear at one level. Generally nectaries are not included in floral diagrams because, while important, nectary location is not used in taxonomic identification of plants. Note the standardized symbols that are used to represent the floral parts in the basic floral diagram illustration to the right. This flower has three petals, three sepals, six stamens, and a pistil made of three united carpels.

1. Select a fully open flower and observe it from above.
   ↪ Draw a top view and record it on the Orientation of Floral Parts sketch sheet (page 76).
   ↪ From this same perspective, create a second drawing that is a floral diagram (cross section) on the same sketch sheet.

2. A **longitudinal section** of a flower represents a side view of the flower as it would appear if it were cut vertically. The longitudinal section can show flower parts and their attachment locations on the receptacle, or central floral axis.
   ↪ Viewing the Fast Plant flower from the side, create a longitudinal section of the flower on the sketch sheet.

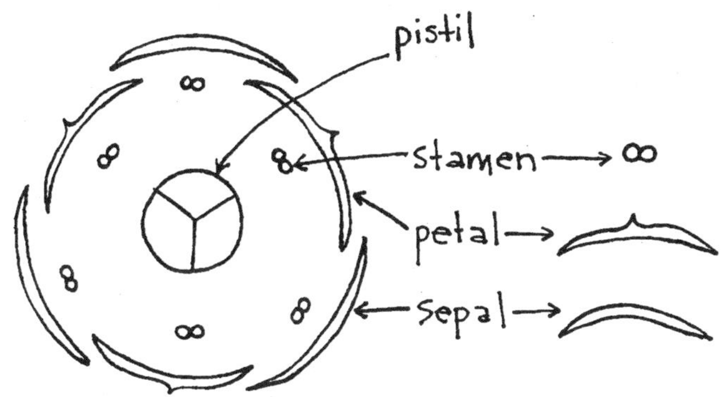

EXAMPLE OF A FLORAL DIAGRAM

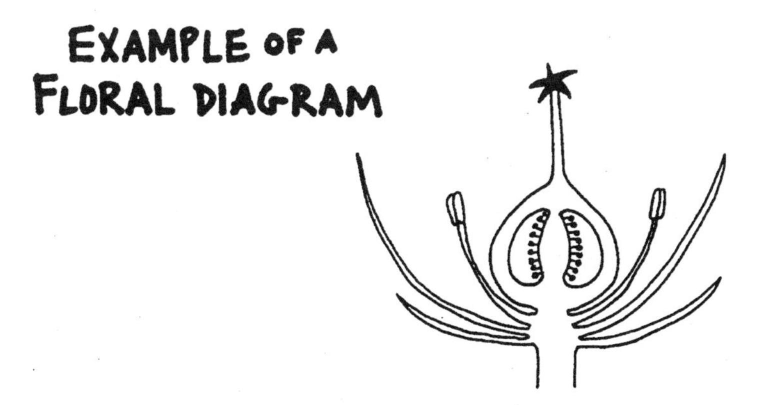

A longitudinal section of the same flower.

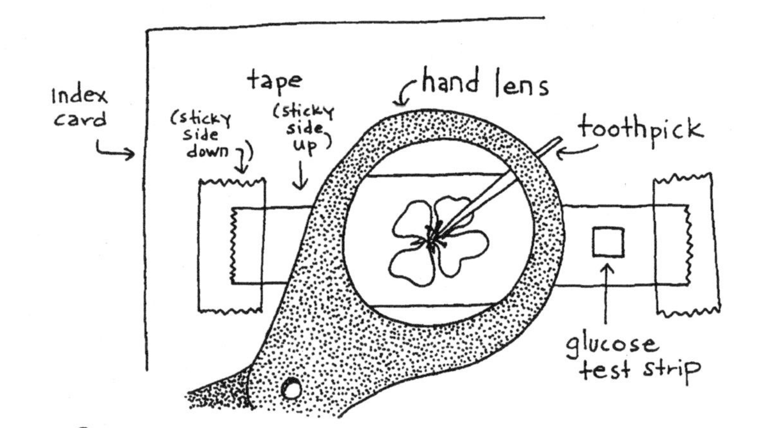

HOW SWEET IT IS

Food for Thought...

The volume and concentration of nectar produced by plants varies from one individual to the next. Researchers have recently determined that, in some plants, production of nectar may be heritable both in terms of total volume and concentration. In addition to the genetic factor, environmental factors can affect nectar production. One researcher has even found that in *Silene latifolia*, a plant with separate male and female flowers, nectar production varied between the sexes!

Nectar can be extracted and measured by touching the base of a nectary with a glass 5-microliter microcapillary tube. Capillary action pulls up available nectar from the flower. The length of the column of nectar can be measured to determine the volume.

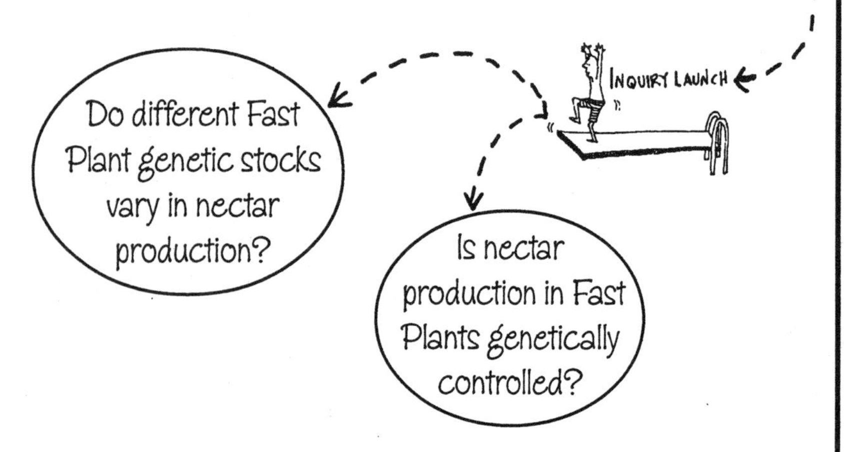

ACTIVITIES, continued

Activity 5
Go for the Glucose

Nectaries are less distinct than other parts of the flower. They can be recognized by the presence of varying quantities of sweet, liquid nectar. With a glucose reagent strip, the presence and concentration of nectar can be easily tested. Generally used to detect sugar levels in urine, a glucose reagent strip reacts with nectar to produce a spectrum of color changes depending on the glucose concentration of the solution.

At the base of the two short anthers in the Fast Plants flowers are two dark green, glistening nectaries producing nectar. Can you prove that the flower really produces nectar? Follow the procedure and go for the glucose!

Procedure

1. Tape an 8-cm piece of tape, sticky side out, to a white index card.

2. Place a Fast Plants flower, which is undergoing anthesis, at one end of the tape and a square of glucose reagent strip at the other end.

3. Moisten the test strip with a very small drop of water using the clean, flat end of a toothpick.

4. While looking through a magnifier, use a toothpick to spread open the flower parts and gently press them to the sticky tape. Note the green nectaries and glistening nectar at the base of the pistil.

5. Probe the nectaries with the clean, pointed end of the toothpick, then touch the tip of the toothpick to the moist test strip. Is there a color change?

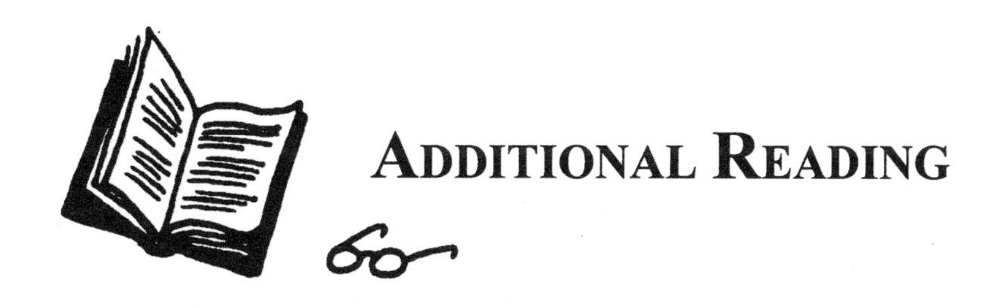

Additional Reading

M. P. Buckles, *The Flowers Around Us: A Photographic Essay on Their Reproductive Structures*, University of Missouri Press, Columbia, 1985.

R. B. Finnell (editor), "The Flower Issue," *Natural History* **108**, 1999. A series of articles focused on the biology of flowers.

P. Leins & C. Erbar, "Floral Developmental Studies," *International Journal of Plant Science* **158**:S3–S12, 1997.

D. J. Mabberly, *The Plant-Book: A Portable Dictionary of the Vascular Plants*, 2nd ed., Cambridge University Press, New York, 1997.

M. A. McKenna & J. D. Thomson, "A Technique for Sampling and Measuring Small Amounts of Floral Nectar," *Ecology*, **69**:1306–1307, 1998.

E. M. Meyerowitz, "The Genetics of Flower Development," *Scientific American* (November), 56–65, 1994. By altering genes for flowering, this researcher has been able to make great strides in understanding how flowers work.

L. Schiebinger, "The Loves of the Plants," *Scientific American* (February), 110–115, 1996. An interesting anthropological story of how the great plant taxonomist Carl Linnaeus used human sexuality concepts of his day (the 18th century) to develop the plant classification system still used today.

A. Zwinger, "Drawing on Experience," in P. Sauer (ed.) *Finding Home,* Beacon Press, Boston, 1992.

Overview

A sequence of five short activities guide students in an exploration of flowers and flowering. In Activities 1 and 2, students study flowers in the context of the whole plant by examining timing of flowering and patterns of flower location on a plant. In Activities 3, 4, and 5, students dissect the flower in order to observe floral parts, their relative orientation to one another, and their function.

Data collected by students on their plants can become part of a class data set, which can then be analyzed, plotted, and displayed. Through this process students may gain a better understanding of the normal variation within a population of Fast Plants.

These activities can be done singly or as a group and the sequence can be altered. Each observation contains launching points for further student-led inquiry.

Objectives

By participating in this activity students will:

- understand that flowering is a key stage of sexual reproduction;
- observe, measure, and analyze variation in flowering among individuals in a population of plants;
- understand that growth in plants represents an ordered sequence of developmental events that vary between individuals of a population;
- understand where and how ovules and pollen originate (male and female gamete formation);
- explore the parts of the flower and the role that each part plays in reproduction; and
- observe the reproductive tissues of plants, including pollen and stigma, under magnification.

Time Required

For All Activities

- Seed must be planted approximately 13–14 days* before Activity 1; 15–17 days* before Activities 2–5 (see calendar).
- 5 minutes are required for thinning plants at 3–5 *das* and 6–8 *das* (see calendar).

Activity 1:
The First Flowers"
- 5 minutes for observation and recording data for 2–3 days surrounding first flowering date.

Activity 2:
The Flower Spiral
- 15–20 minutes for observation and recording data.

Activity 3:
What's in a Flower?
- 50 minutes for dissection.

Activity 4:
Orientation of Flower Parts
- 20–30 minutes for observation and recording data.

Activity 5:
Go for the Glucose
- 20–30 minutes for dissection.

* Days to flowering will vary depending on environmental conditions in your classroom.

Materials

Each student will need:

Activity 1

- 1 flowering Fast Plant (13–14 *das*) in wickpot*
- hand lens or microscope
- Student Data Sheet, *The First Flowers*

Activity 2

- 1 flowering Fast Plant (15–17 *das*) in wickpot*

Activity 3

- 1 flowering Fast Plant (15–17 *das*) in wickpot*
- hand lens or microscope
- fine-tipped forceps, dissecting needle or toothpick
- white index card with about 8-cm piece of tape, taped to card, sticky side out.

Activity 4

- 1 flowering Fast Plant (15–17 *das*) in wickpot*
- hand lens or microscope

Activity 5

- 1 flowering Fast Plant (15–17 *das*) in film can wickpot*
- hand lens or microscope
- forceps
- white index card with about 8-cm piece of tape, taped to card, sticky side out.
- toothpick
- Glucose Diastix® reagent strip for urinalysis, available from local drugstore (one strip is sufficient for 8–10 tests).

* See Appendix 1 (page 113) for growing instructions.

Classroom Management Tips

With three students working as a team with one deli reservoir, each student can be responsible for two plants in a subpopulation of six. A spare seventh film can may be planted. Students sow four seeds in each film can wickpot and place them in an environment conducive to germination, growth and emergence. At 3–5 *das*, students thin to two plants/film can. At 6–8 *das,* students thin to one plant/film can. Approximately 13–14 days after sowing, plants will begin flowering.

FAST PLANTS: FLOWERING ACTIVITIES CALENDAR

MONDAY	TUESDAY	WEDNESDAY	THURSDAY	FRIDAY
	Plant	1 (das)	2	3
← - - - - 6	Thin to 1 plant/wickpot 7	- - - - → 8	9	10
← - - Record Opening of 1st Flower 13	- - - → ← - - 14	Record Opening of Flowers #2-5 ← - - DISSECTION - - 15	- - - 16	- - - → 17

The First Flowers

Individual Data Sheet

When do flowers open on a Fast Plant?

Observation Time	Observation Date	*das*	Number of open flowers	Number of open flowers in last 24 hours

Assuming that Fast Plant flowers open at regular time intervals, what is the average time between the opening of successive flowers? Show your calculations below.

Flower Spiral

Individual Data Sheet

What is the pattern of flowers on the plant stem?

1. Mark the film can with tape such that the first flower is lined up with the 0°/360° mark. Always orient film can such that the tape is lined up with the 0°/360° mark.

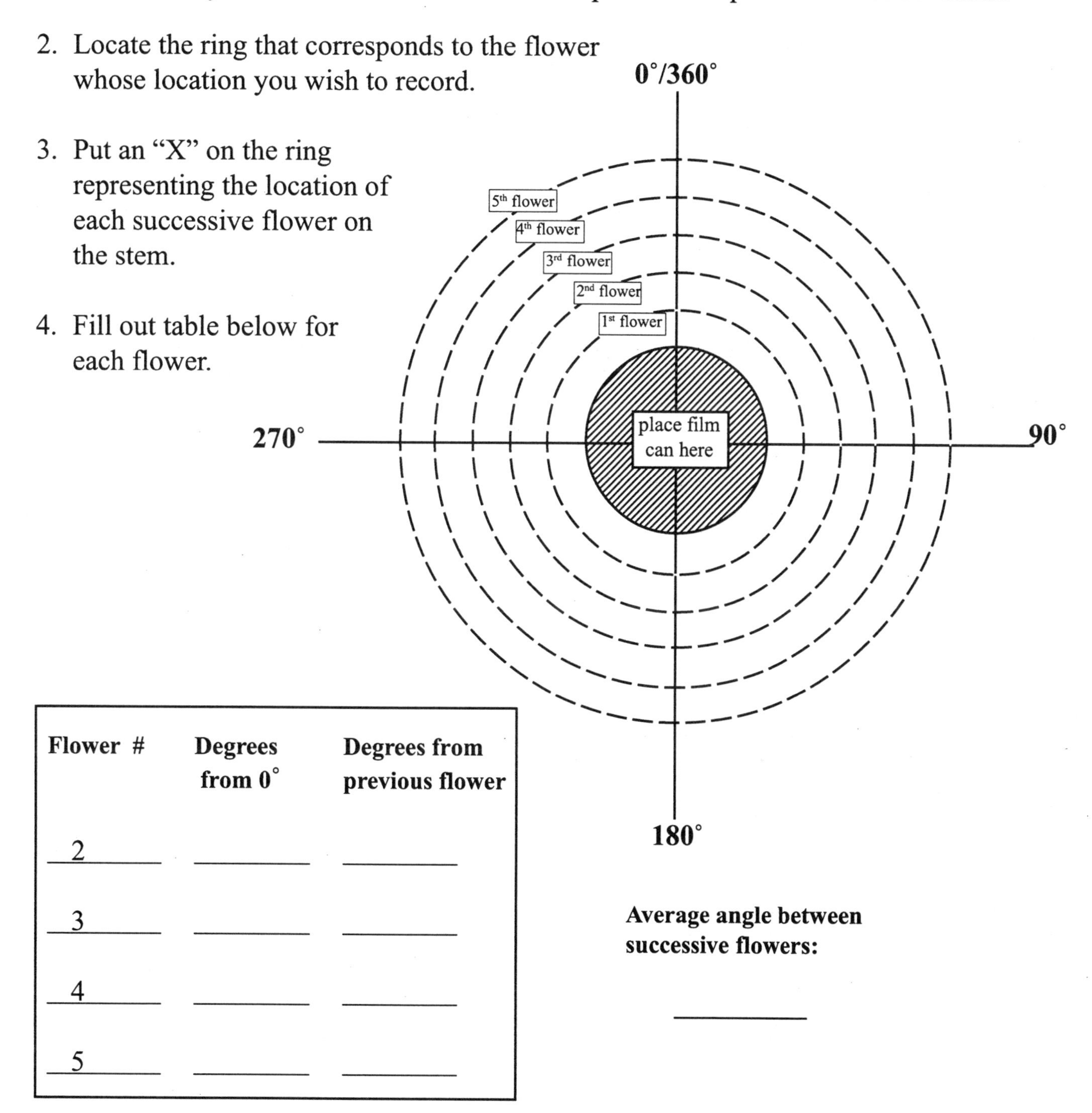

2. Locate the ring that corresponds to the flower whose location you wish to record.

3. Put an "X" on the ring representing the location of each successive flower on the stem.

4. Fill out table below for each flower.

Flower #	Degrees from 0°	Degrees from previous flower
2	________	________
3	________	________
4	________	________
5	________	________

Average angle between successive flowers:

____

FLOWER SPIRAL

Class Data Sheet

Plant #	Degrees between flower 1 and flower 2	Degrees between flower 2 and flower 3	Degrees between flower 3 and flower 4	Degrees between flower 4 and flower 5	Average angle between successive flowers
1					
2					
3					
4					
5					
6					
7					
8					
9					
10					
11					
12					
13					
14					
15					
16					
17					
18					
19					
20					
21					
22					
23					
24					
25					

Flower Spiral

Analysis of Class Data

	Degrees between flower 1 and flower 2	Degrees between flower 2 and flower 3	Degrees between flower 3 and flower 4	Degrees between flower 4 and flower 5	Average angle between successive flowers
Class mean	________	________	________	________	________
Class median	________	________	________	________	________
Class range	________	________	________	________	________
Class standard deviation	________	________	________	________	________

Orientation of Floral Parts

Sketch Sheet

How are the parts of a flower arranged?

Sketches of a Fast Plant Flower from Three Perspectives

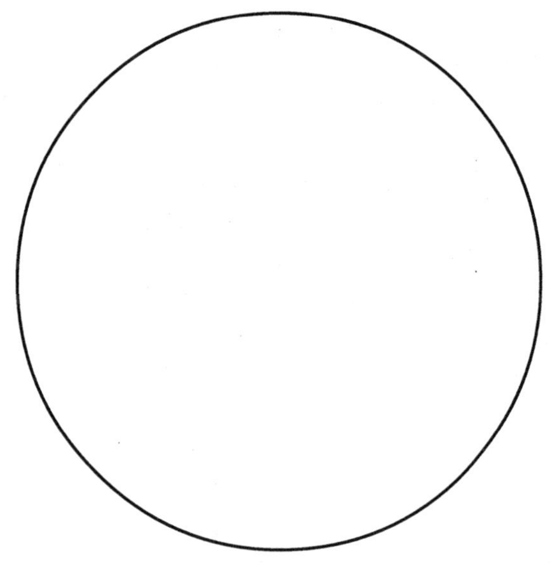

Top-Down View

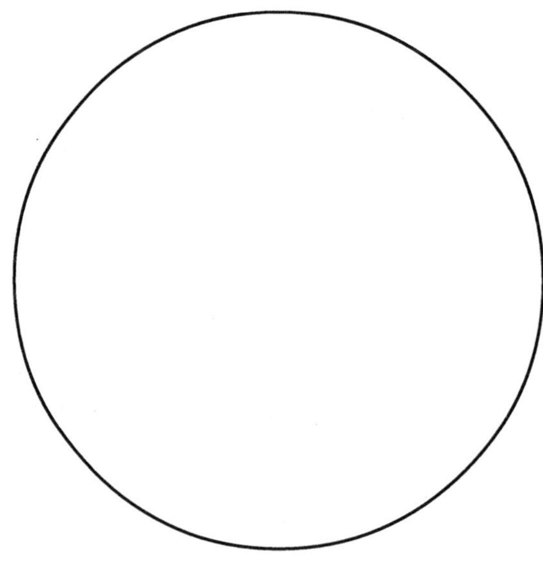

Floral Diagram
(Top-down, cross section)

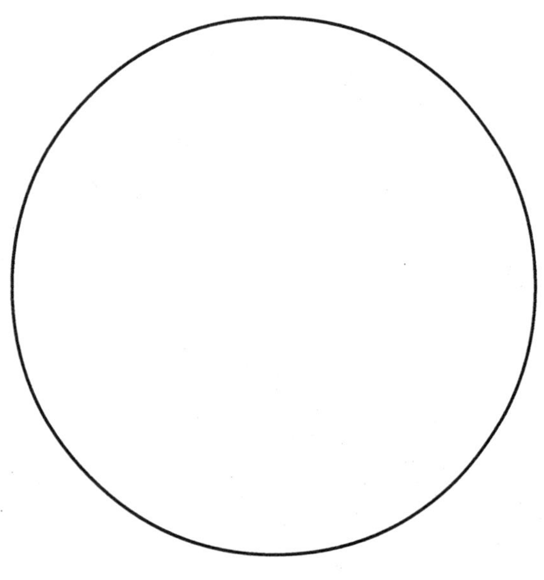

Longitudinal Section
(side view, sliced vertically)

What are the advantages and disadvantages of each of the three ways of drawing a flower? What information is conveyed in some views that is not conveyed in others?

CHAPTER 4

POLLINATION

How do bees and brassicas relate?

IT TAKES TWO

Symbiosis is the close association of two or more dissimilar organisms. Some symbiotic relationships, typically disease-causing, are *parasitic* (beneficial to one organism and detrimental to the other). *Mutualistic* symbiotic relationships (beneficial to both organisms), such as a flower and its pollinator, are commonly found in nature. Under close examination, each mutualistic relationship stands out as an example of evolutionary complexity.

When two or more symbiotic species evolve in response to each other, they are said to *coevolve*. The coevolution of bees and certain flowers, including brassicas, each dependent upon the other for survival, is such a relationship. ⊠

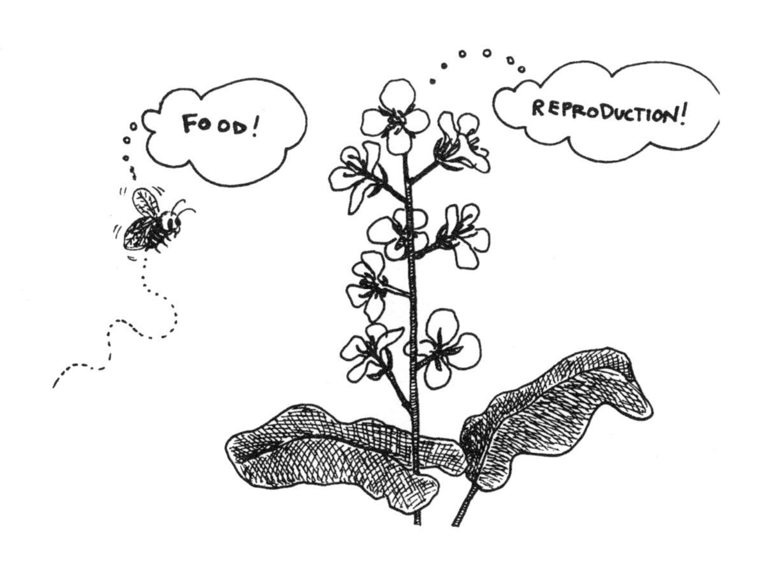

BACKGROUND

Insect Pollination: A Symbiotic Relationship

A mobile insect needs food, and an immobile plant needs another plant's pollen for fertilization; together they meet one another's needs. This is an example of one of the most fascinating relationships in the natural world—insect pollination.

Pollination is the process of mating in plants whereby pollen grains developed in the anthers are transferred to the stigma. The pollen grains then germinate, forming pollen tubes that carry the sperm to the eggs that lie within the pistil. Not all flowering plants rely on animal pollinators to transfer pollen from anthers to stigmas. Some plants self-pollinate within the unopened flower bud, while wind or water moves the pollen in other plant species.

The evolution of flowering plants and their pollinators has resulted in a diversity of mechanisms that deliver pollen to the stigma. Flowers may be pollinated by a specific bee, bird, bat, or beetle, or by several different organisms. Flowers often attract

BRASSICA FLOWER

HONEY BEE

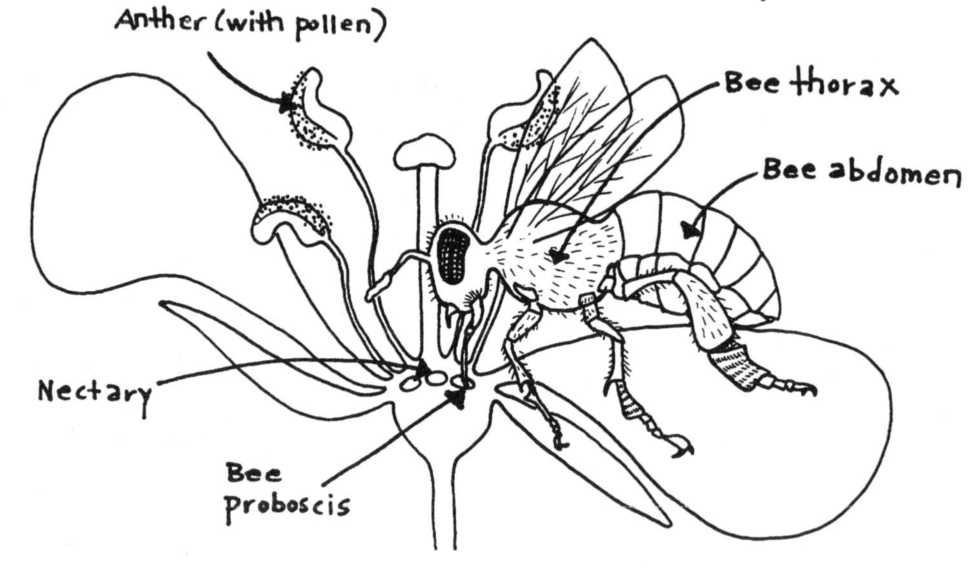

BEE DIVERSITY

There are over 20,000 different species of bees in the world. Though they are very diverse in behavior and appearance, they all have *setae* (feather-like hairs) that hold pollen and they all feed on nectar and pollen. Bees depend on many different kinds of flowers. Given the diversity of flowering plants, it is clear why there is a great diversity of bees. Unlike the familiar honeybee, the vast majority of bees are solitary and do not live in a colony. Some bees spend the winter alone, and then form colonies during the summer.

Farmers know that pollination can be the limiting factor in the yields of many crops. Because of this, it is common to bring in and release commercially bred bees into a field (or greenhouse) of a flowering crop. Although honeybees are most commonly used for this purpose, another type (such as bumblebees or leaf cutter bees) is brought in if they cannot pollinate the flowers of a particular crop. ⊠

BACKGROUND, continued

certain pollinators by the color, fragrance, or structure they possess. Insects, for their part, often have specialized sensory apparatus, body parts, and behaviors that allow them to successfully collect pollen and nectar from the flowers they visit.

The Bee and the Brassica

In brassicas, bees and other insects distribute pollen. Brassica pollen is heavy and sticky—unable to be easily wind-borne. Bees are marvelously coevolved pollen *vectors* (transferring devices) for brassicas.

Bees depend on flowers for their survival. Sugars in the nectar provide carbohydrates to power flight and life activities. Pollen is the primary source of proteins, fats, vitamins, and minerals to build muscular, glandular, and skeletal tissues in bee larvae.

BACKGROUND, continued

Bees are members of the insect family Apidae, which are unique in that their bodies are covered with *setae* (feather-like hairs). The bright yellow petals of brassica flowers act as both beacons and landing pads for the bees, attracting them to the flower and guiding them to the nectaries. The bee drives its head deep into the flower to reach the sweet nectar secreted by the nectaries, brushing against the anthers and stigma in the process. Quantities of pollen are entrapped in its body hairs.

As the bees work from plant to plant, pollen on the setae is carried from flower to flower. The transfer of pollen from an anther to a stigma is known as *pollination*. When pollen is transferred from one plant to another, the process is called *cross-pollination*.

INQUIRY LAUNCH

How does the varying length of the stamens and the orientation of the anthers relate to the position of the bee as it visits a brassica flower?

KNOW THYSELF

For Fast Plants and many other brassicas, the act of pollination does not insure fertilization and seed formation. *Brassica rapa* and other species contain biochemical recognition compounds called *glycoproteins* that are unique to each plant. These compounds enable the plant to recognize "self," resulting in the rejection of the plant's own pollen. This genetically-controlled prevention of fertilization with "self" pollen is called *self-incompatibility*. Only "non-self" pollen is able to germinate and affect fertilization.

In order for pollen germination and fertilization to occur, pollen must travel from one brassica plant to the stigma of a different brassica plant in the process of cross-pollination. Bees take care of this problem naturally as they move from plant to plant in search of nectar and pollen. Fast Plants are cross-compatible and relatively self-incompatible.

Our own bodies also contain glycoprotein recognition compounds. These glycoproteins are important in identifying tissue compatibility in organ transplants and are critical to getting a match in which the tissue glycoproteins of the donor are the same as in the recipient. Unlike self-incompatibility in plants, in human tissue *compatibility* is triggered by similar self-like glycoproteins.

HONEY BEE LEGS

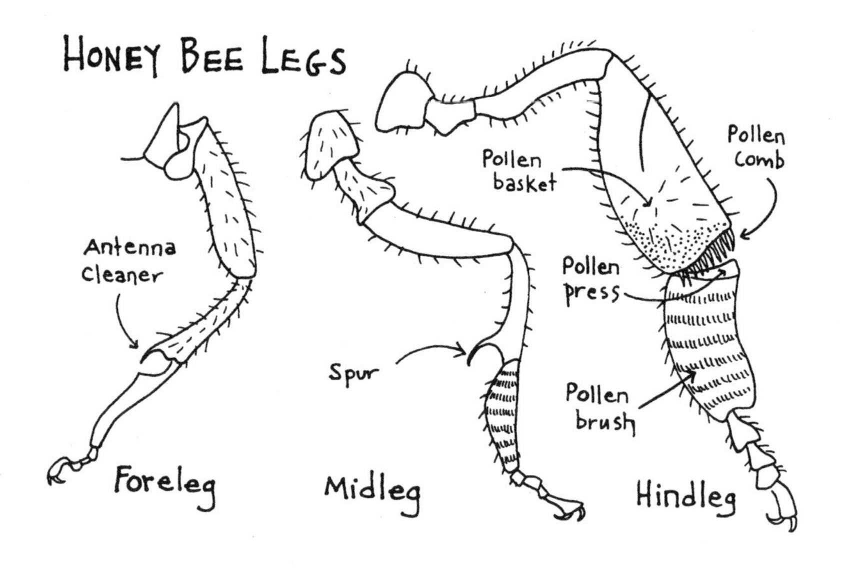

BUSY BEES

A colony of honey bees (*Apis mellifera*) will collect 44 to 110 pounds of pollen in a season. The bee's three pairs of legs are adapted in several ways for efficient pollen collection (see illustration). First, the legs have evolved to comb pollen from the bee hairs and pack it into a single spot called the *pollen basket* for transport to the hive. Using the large flat *pollen brushes* on the hindlegs, the bee then removes the pollen from its head, thorax, and forelegs. The pollen is raked off the brushes by the *pollen combs*, mixed with a little nectar to make it stick, and packed into the baskets by the *pollen press*. Finally, each foreleg is equipped with an *antenna cleaner*, a deep notch with a row of small spines, which is used to brush pollen from the antennae.

Beyond pollination, what might be the specialized roles of other bee body parts?

BACKGROUND, continued

As the bee forages from plant to plant, cross-pollination occurs and genetic information is spread widely within the population. Pollen grains are deposited on the sticky surface of each stigma and each compatible pollen grain sends a tube through the style to an ovule. Within 12 hours of pollination, some of the pollen grains will then complete fertilization. Within two days of fertilization, the flower's petals begin to drop and the pistil begins to elongate to form a pod as the seeds develop inside.

ACTIVITIES

Activity 1
Bees and Beesticks

As your Fast Plants come into flower, be prepared to pollinate. Pollination is the prelude to a new generation that starts with fertilization. Before jumping into pollinating, this activity provides an opportunity to consider the concepts associated with the coevolutionary relationships between flowers and their pollinators.

Procedure

1. Observe the anatomy of your bee with a hand lens, focusing on the legs and proboscis. Identify the pollen basket, antenna cleaner, pollen brushes, pollen combs, and pollen press (see illustration).

?? What parts of the bee's anatomy ensure that its needs are met in pollination?

↪ On the Bees and Beesticks sketch sheet (page 88), sketch the bee, circling the parts of its anatomy that are involved in pollination.

2. Carefully examine your Fast Plant flower.

?? What parts of the flower's anatomy ensure that its needs are met in pollination?

↪ On the same sketch sheet, draw the flower, circling the parts of its anatomy that are involved in pollination.

?? How is the anatomy of the bee related to the anatomy of the flower?

↪ Draw lines between portions of the bee and flower anatomy that are related in pollination and nectar foraging. Briefly describe the nature of the relationship.

3. Because bees are well suited for pollinating Fast Plants, you will pollinate your plants with them! However, as live bees are rather hard to control in the classroom, you will be making a beestick for pollination rather than relying on the living organism. Carefully remove the legs, head, wings, and abdomen of the dried bee, leaving the fuzzy thorax. (These bees died in the hive; in making a beestick, you are recycling them and using them once again for pollination!)

4. Place a drop of fast-drying glue on the tip of a toothpick. Carefully push the toothpick into the bottom of the bee's thorax, near the leg joints. Let the beesticks dry for at least several hours.

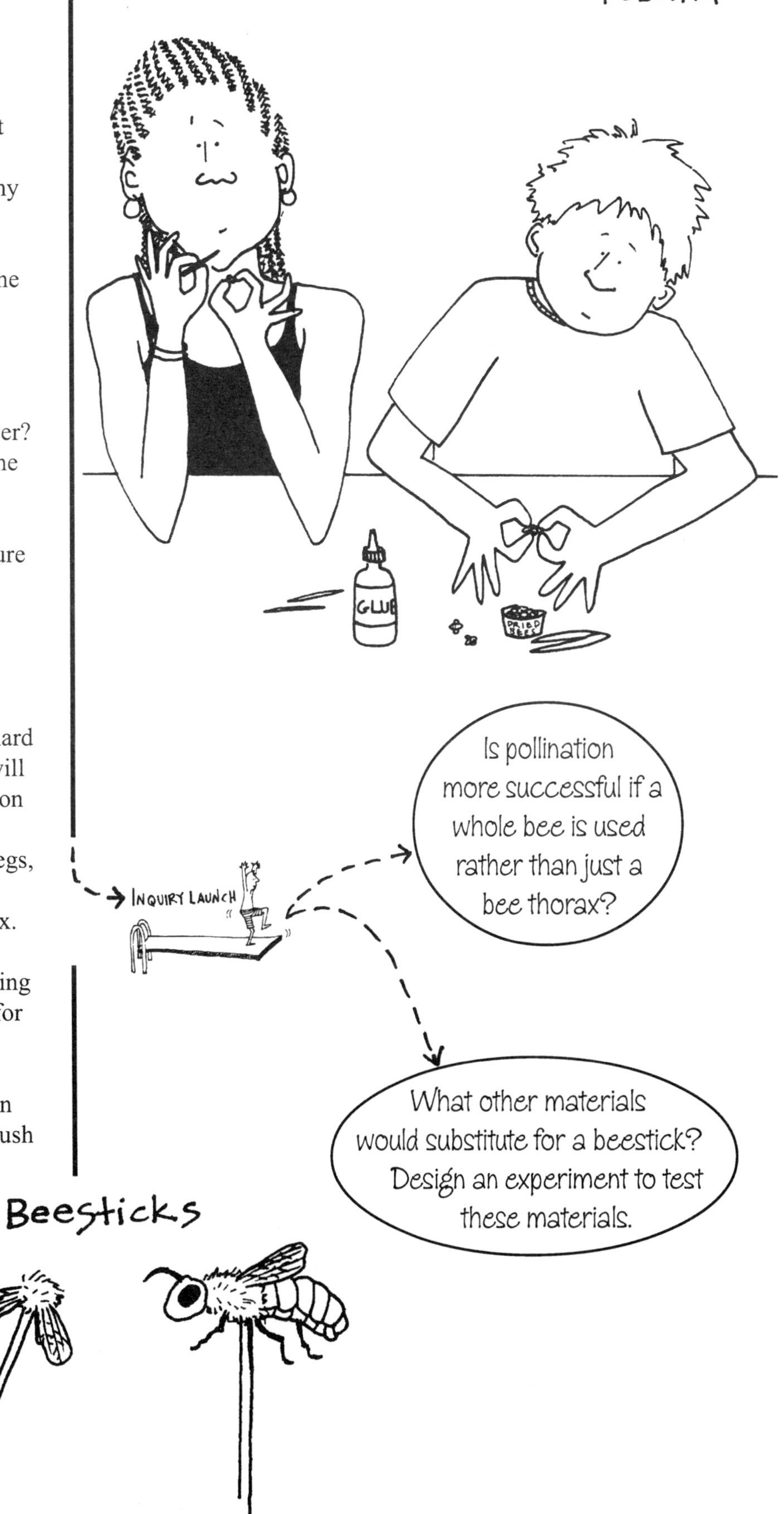

ACTIVITIES, continued

Activity 2
It's Pollination Time

1. Sometime between **Day 14** and **Day 16**, when five or more flowers are open on each of your plants, pollinate them. Holding a flower in one hand, gently roll the beestick back and forth over the anthers until yellow pollen can be observed on the hairs of the bee thorax. Repeat with all open flowers on each of your plants.

2. Moving to another person's plants, repeat the gentle rolling motion over the anthers and the stigma of each pistil, making sure to deposit pollen from your beestick on the stigma of each flower.

3. Before leaving the plants, ensure that each of the flowers has been adequately pollinated by the beestick. Can you see any pollen on the stigmas? Examine the stigma with a hand lens.

4. Observe pollen attached to the bee setae or on stigmas of dissected flowers with a hand lens or higher-powered microscope. Alternatively, a beestick with pollen can be rolled over the sticky tape on a Scale Strip and examined.

   Pollination should continue every day or two for 3–4 days after the first flowers open on the plants. The completion of pollination sets the floral clock at "0 *dap*" (days after pollination).

Activity 3
Pollen Germination
Pollen from Fast Plants can be stimulated to germinate in a sucrose–salt solution, maintained as a hanging drop slide. Within 30–40 minutes, young pollen tubes can be observed under magnification, growing from the pollen grains.

Procedure
1. Place 6 drops of salt solution followed by 2 drops of sucrose solution into the "well" of a clean, clear Fuji® film can lid. Mix the liquids by sucking them into a pipette and squirting them back into the film can lid several times.

2. Place one drop of the mixture in the center of a glass microscope slide.

3. Pluck a stamen from a fresh Fast Plant flower and touch the anther to the drop, dislodging a small amount of pollen on the slide. A single stamen can be used for several slides.

4. Carefully invert the slide so that the drop hangs from the underside and place the slide over the "well" of the film can lid, which should contain some residue of solution or water to maintain a humid environment around the hanging drop.

5. After 30–40 minutes, short pollen tubes (early in development) will be visible at a magnification of 10X or greater. The full-length pollen tubes will form in 60–90 minutes. It is possible to wait 24 hours and observe the tubes in the next class period.

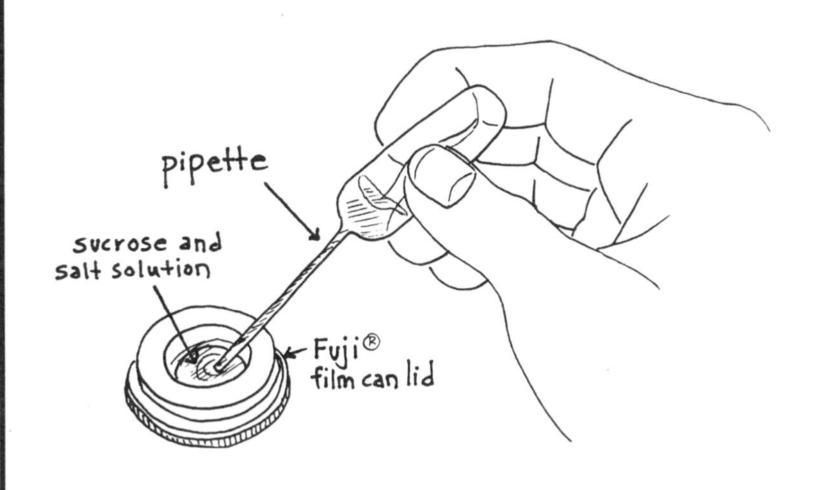

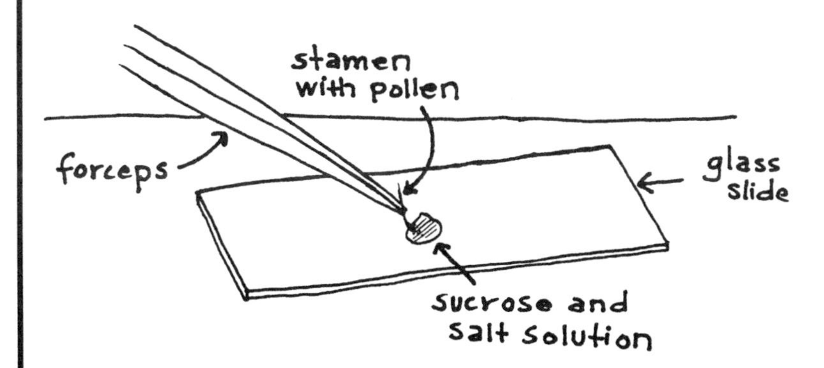

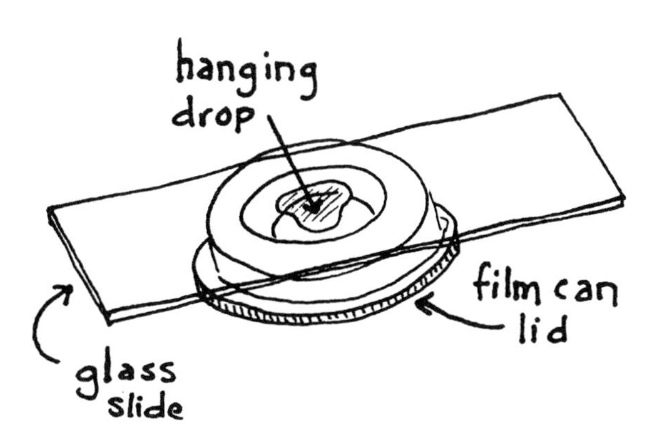

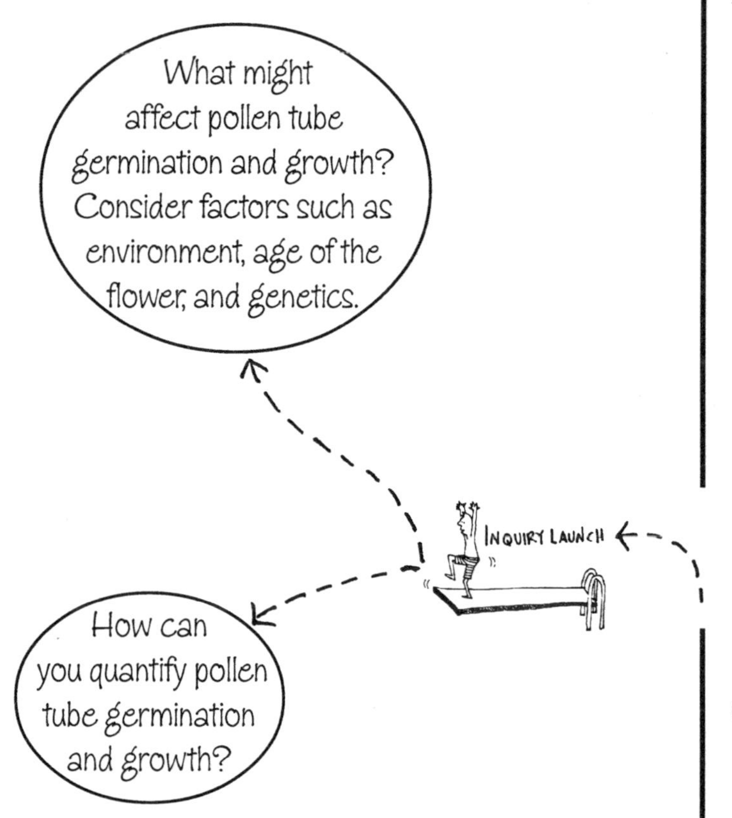

ACTIVITIES, continued

The tubes may be dry at this point, but they will still be visible. To prevent the hanging drops from drying out, keep the slide inverted over the film can lid and place the entire apparatus on a piece of moist paper towelling in a petri dish. Tape the petri dish shut and place into a ziplock bag.

To observe the pollen tubes, carefully turn the slide over, so that the drop is again on top. Place the slide on the stage of a compound microscope and observe. If you examine the drop at greater than 100X magnification, place a cover slip over the drop to avoid damage to the microscope.

↪ On the Pollen Germination sketch sheet (page 89), sketch the pollen tubes as they develop.

Additional Reading

L. Gamlin, "The Big Sneeze," *New Scientist* (June), 37–41, 1990.

C. A. Kearns & D. W. Inouye, "Pollinators, Flowering Plants, and Conservation Biology," *BioScience* **47**:297–306, 1997. Reviews the dependence by plants on pollinators and the effects on plant communities from losses in insect pollinators.

K. H. Desmond, "Chronicle of the Lustful Plants," *New Scientist* (April), 57–61, 1989.

R. Mestel, "Not Doing It, Plant Style," *Discovery* (January), 1995. An article on plant self-incompatibility.

D. C. Robacker, B. J. D. Meeuse, & E. H. Erickson, "Floral Aroma: How Far Will Plants Go to Attract Pollinators?" *BioScience* **38**:390–397, 1988. Illustrates the nature of floral aroma communication systems through three case studies.

Overview

These activities will provide students with insight into the significance of pollination and the complex web of interactions that accomplishes it. In Activity 1, students carefully examine the structure and function of a honeybee and a Fast Plant flower. They draw connections between the two to better understand the symbiotic relationship of pollination. After pollinating their Fast Plants with beesticks in Activity 2, students observe the development of a pollen tube in Activity 3. This chapter sets the stage for the exploration of the intricacies of fertilization in the next chapter.

Objectives

By participating in this unit students will:

- understand flowering as the sexually-mature stage of plant development;
- observe the reproductive tissues of plants, including pollen and stigma, under magnification;
- understand the interdependent coevolutionary relationship of bees and brassicas;
- observe pollen tube development; and
- pollinate their Fast Plants using a beestick, thereby setting the stage for future developmental events.

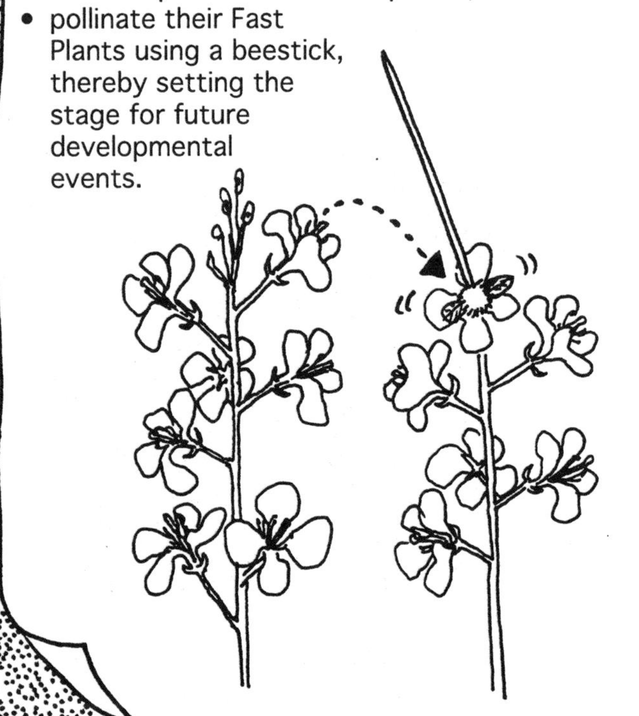

Time Required

Activity 1:
Bees and Beesticks

- 50 minutes.

Activity 2:
It's Pollination Time

- 10 minutes per day for 3–4 days.

Activity 3:
Pollen Germination

- 10–20 minutes to set up, observations can be made 30–90 minutes later or can wait until next class period.

Preparation of Solutions

To prepare 1.2 M sucrose solution, dissolve 41 grams of sucrose (table sugar) in enough distilled water to make a final volume of 100 ml. This is equivalent to one level tablespoon of sugar in the volume that will fill one 35-mm film can.

Prepare a mineral salt solution by adding the following quantities of salts to distilled water to bring the final volume to one liter:

- 0.417 g $Ca(NO_3)_2$ (calcium nitrate)
- 0.200 g H_3BO_3 (boric acid)
- 0.101 g KNO_3 (potassium nitrate)
- 0.217 g $MgSO_4 \cdot 7H_2O$ (magnesium sulfate)

Keep this solution refrigerated to avoid bacterial and/or fungal growth.

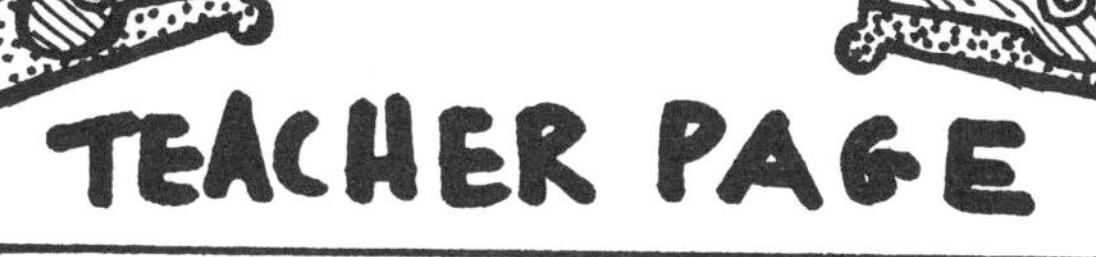

Materials

Each student will need:

Activity 1

- 1 dried bee
- 1 Fast Plant flower
- round toothpicks
- glue (e.g., Duco® Cement)
- hand lens
- forceps

Activity 2

- 1 beestick
- 2 flowering Fast Plants (Day 14 to 16) in a wickpot*
- 2-cm wide clear tape or Scale Strip (page 135)
- hand lens or microscope

Activity 3

- disposable plastic transfer pipette
- clear Fuji® film can lid
- microscope slide
- 1 stamen (can be used for several students)
- compound microscope
- 1 drop of pollen germination solution described in Preparation of Solutions (page 86)

* See Appendix 1 (page 113) for growing instructions.

Classroom Management Tips

- Beesticks should be made one or two days prior to pollination.
- When making beesticks the bees may be very dry and brittle. Soften the bees by placing them in an airtight container with a little moist paper towel overnight.
- A period of 16 days from the sowing of seed is required for the growth of the Fast Plants and the completion of the activity.
- For the Pollen Germination activity, the age of the flowers is important. Newly opened flowers seem to work better.

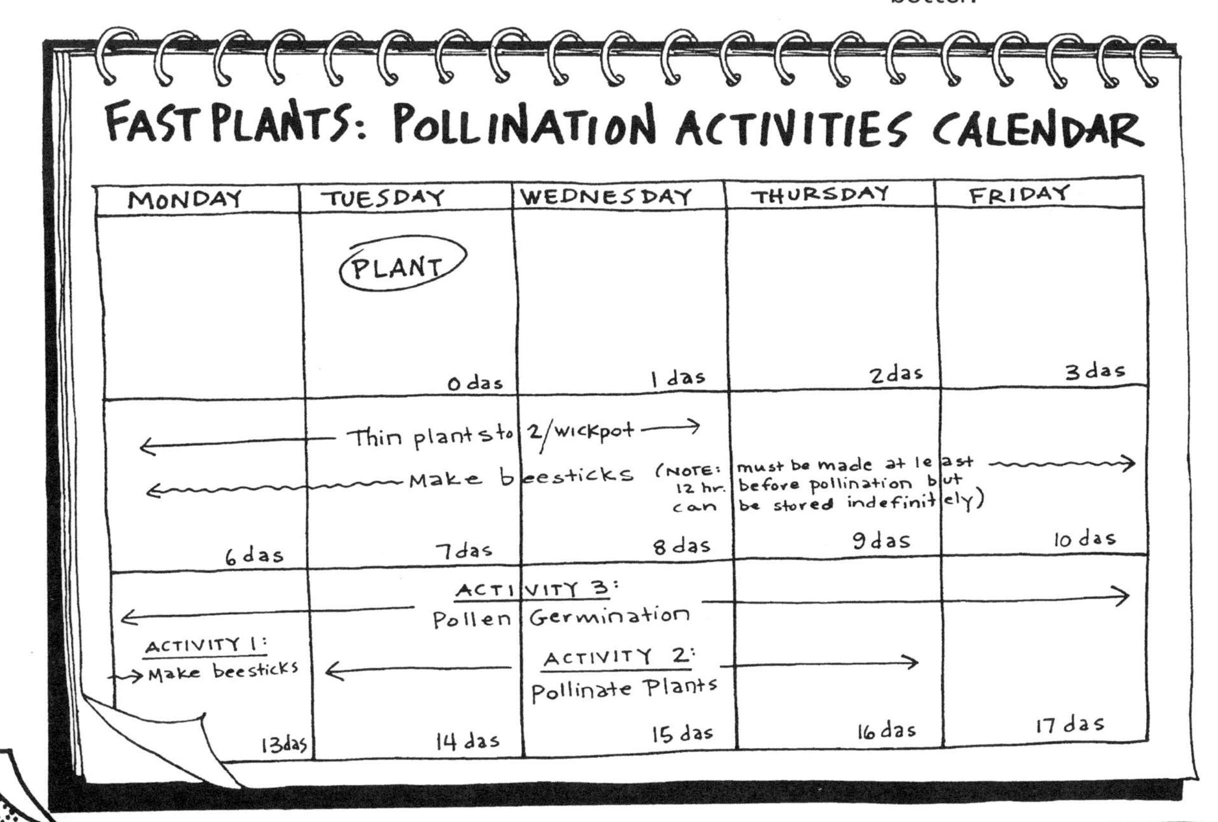

BEES AND BEESTICKS

Sketch Sheet

How is the anatomy of the bee related to the anatomy of the brassica flower?

What is the magnification of your drawing? See Appendix 4 (page 135) for drawing to scale and calculation of magnification.

Sketch of Bee and Brassica

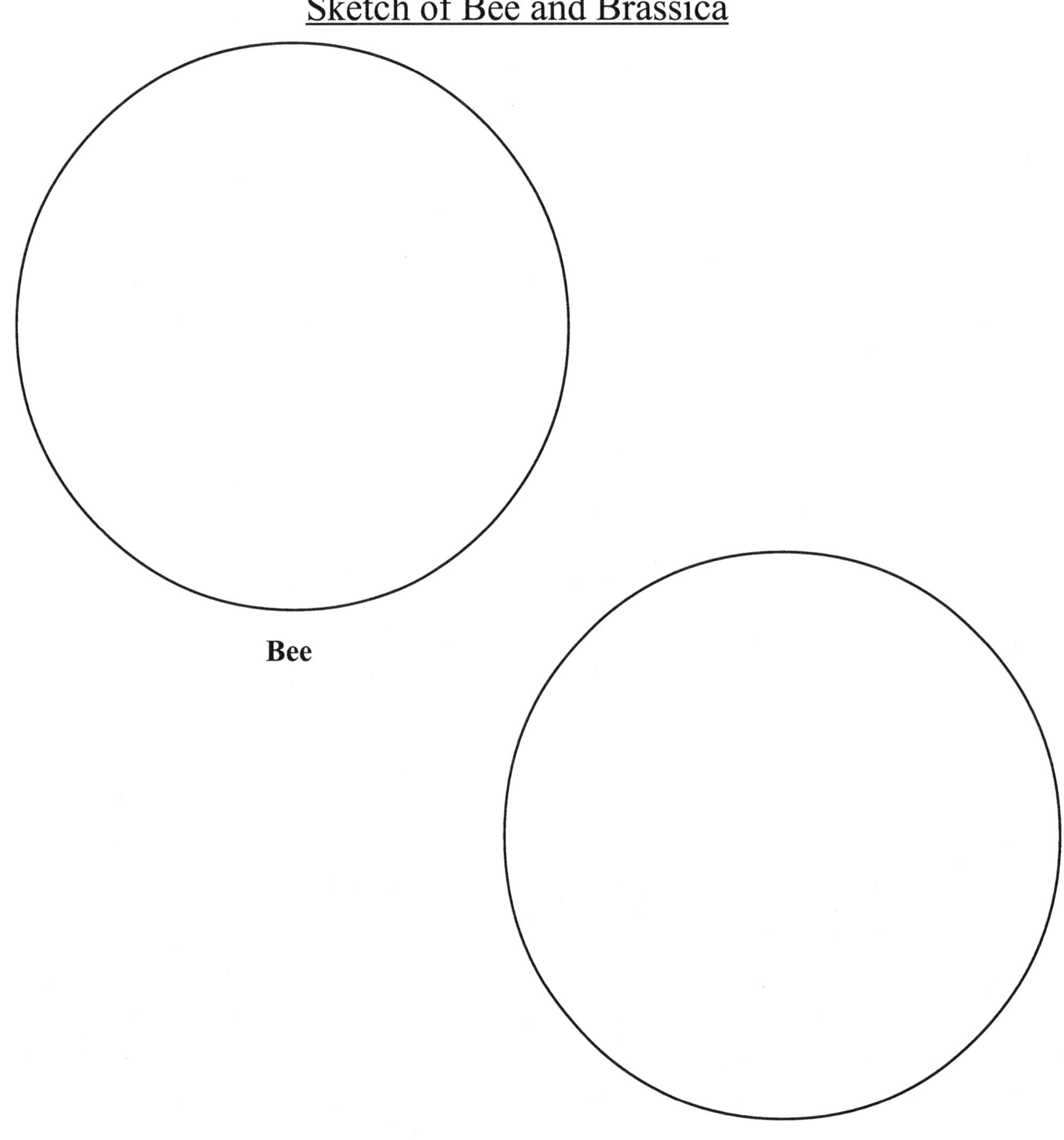

POLLEN GERMINATION

Sketch Sheet

How do pollen tubes germinate and grow?

What is the magnification of your drawing? See Appendix 4 (page 135) for drawing to scale and calculation of magnification.

Sketch of Germinating Pollen Tubes

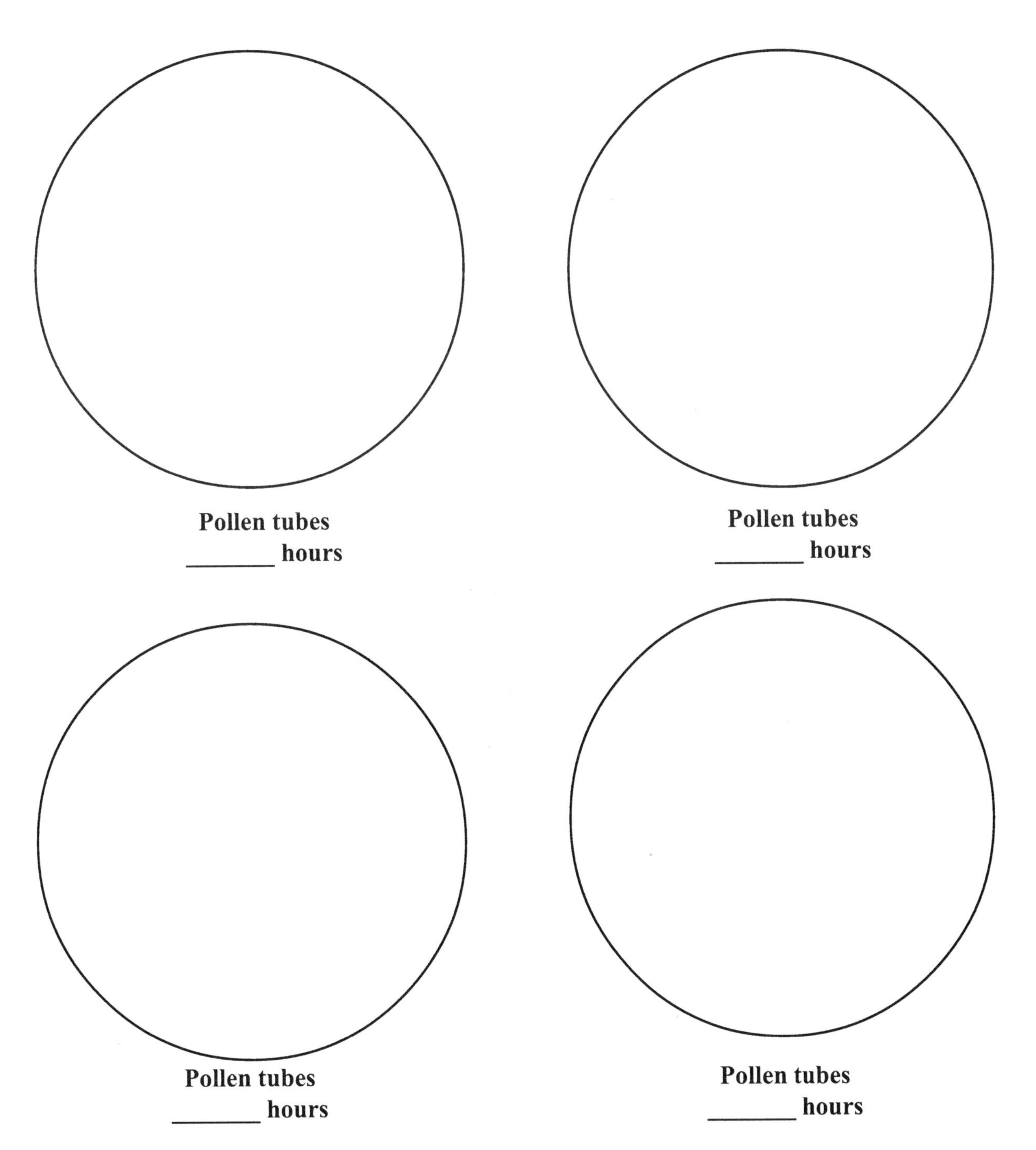

CHAPTER 5

FERTILIZATION TO SEEDS

What happens when sperm meets egg?

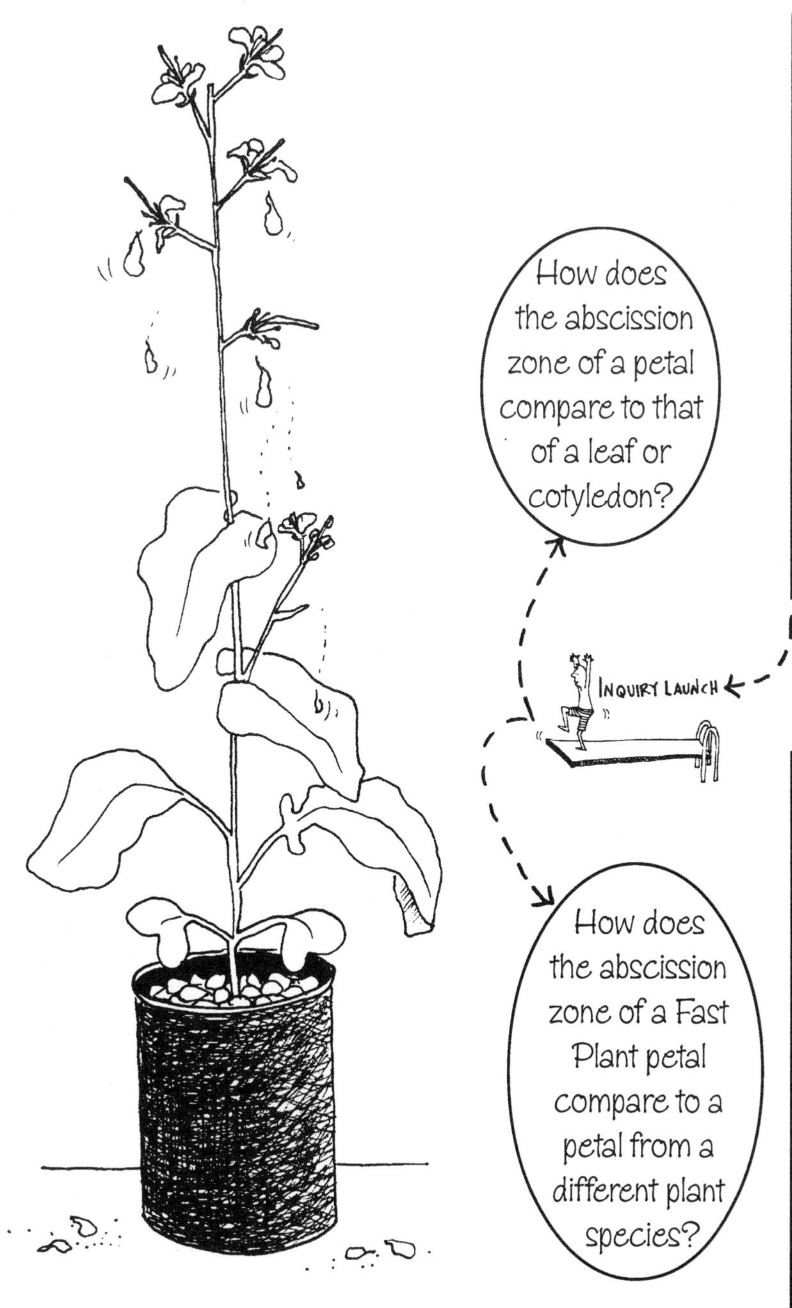

BACKGROUND

Alive or Dead?

What happens between fertilization and seed harvest? Fertilization in sexually reproducing organisms represents the beginning of the next generation. Following fertilization in plants, the most visible event is the rapid wasting of all those flower parts that are no longer needed by the plant. Having completed their functions, nectaries dry up. Sepals, petals, and stamens wither and fall. This shedding of organs from the body of the plant is called *abscission* and is preceded by both chemical and structural changes at the abscission zone. It may appear that life is over, that the plant is dying. Nothing could be further from the truth.

Immediately following fertilization, the pistil and other maternal structures will grow and change functions. Within the ovules, the embryos also grow and differentiate through a series of developmental stages known collectively as *embryogenesis*. Although the plant is not dying, it is certainly changing. And, through these changes, another turn in the spiral of life is beginning.

FERTILIZATION TO SEEDS

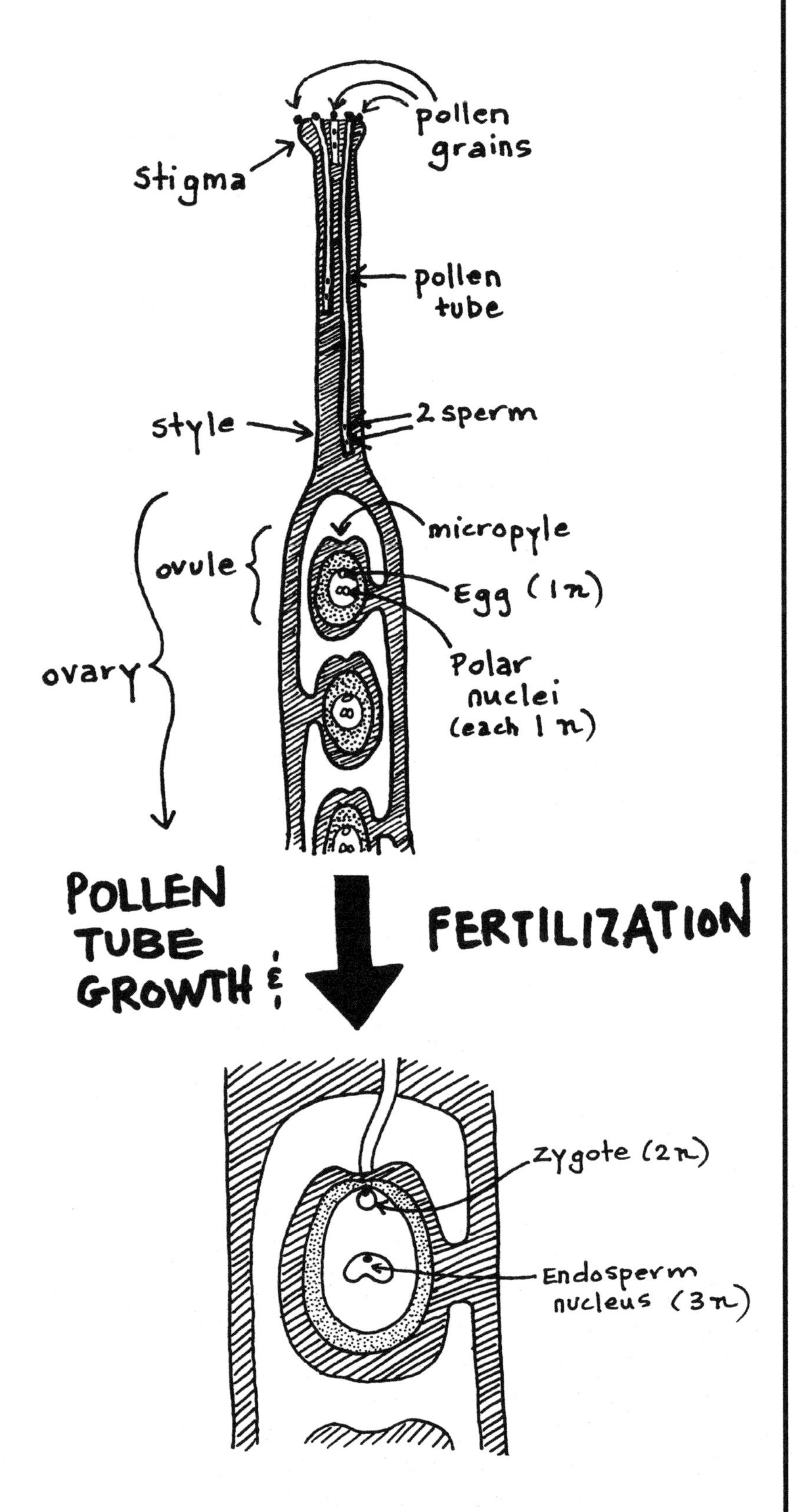

BACKGROUND, continued

Fertilization: 2 Sperm and 1 Egg!?
Look more closely at fertilization in higher plants such as Fast Plants. After pollination, a pollen tube grows from each of the many compatible pollen grains adhering to the stigma. Each tube contains two sperm cells and its own nucleus. Only a few of the hundreds of pollen tubes that enter the ovary cavity will successfully fuse with an ovule, and each ovule is joined to only one tube.

One sperm from the pollen tube then unites with the egg cell nucleus (1n) in the ovule to produce a zygote (2n), which will develop into the embryo. The second sperm unites with the ovule's two polar nuclei (each of which are 1n) to form the endosperm (3n). This process is sometimes referred to as *double fertilization* in that it involves two sperm uniting with two separate nuclei.

After Fertilization: Embryogenesis
Embryogenesis is a highly coordinated sequence of developmental events within the ovule and supporting maternal ovary tissues that ultimately leads to the development of a viable seed and accompanying fruit. The following processes of embryogenesis are responsible for the production and packaging of the next generation (the seed):

- the endosperm is formed;
- the zygote develops into an embryo;
- ovule cells differentiate to produce a seed coat; and
- the ovary wall and related structures develop into a fruit.

After fertilization, and before embryo development, the endosperm and supporting maternal tissues rapidly grow and develop. The triploid (3n) endosperm nucleus that formed during fertilization divides very rapidly and repeatedly to form the nutrient-rich, starchy liquid endosperm. This liquid endosperm bathes the developing embryo, providing it with nutrients. In the latter stages of embryo development of brassicas and other plants, the embryo converts the starchy reserves in the endosperm into lipids that are stored in the embryonic cotyledons. As the embryo matures to a seed, it comes to occupy the space that was filled by the endosperm.

As endosperm formation begins, the first mitotic division of the zygote marks the beginning of embryogenesis. After successful pollination and fertilization, Fast Plant embryos mature into seeds in 20 days.

While the embryo develops, the *integuments* (the walls of each ovule) develop into a seed coat. This coat of maternal tissue protects the new generation until favorable conditions for seed germination are present.

Finally, as the ovule develops into a seed, the ovary wall and other maternal structures in the pistil grow to become the *fruit*. In some plants, this tissue (which surrounds the enlarging seeds) may thicken, differentiate and develop into a fleshy fruit. In other plants, such as with Fast Plants, it may dry down into a *pod*. In addition to protecting the developing ovules, the fruit often serves as a means of seed dispersal.

TOGETHER: MOTHER and CHILD

Food for Thought...

Embryogenesis involves a complex series of overlapping processes that are closely coordinated. Fertilization seems to initiate this coordination. It synchronizes formation of the endosperm and of the embryo: without one, the other will not proceed. Energy and resources will be put toward endosperm development only if it is associated with an embryo in need of nourishment.

Recent studies utilizing both tissue culture techniques and genetic analysis indicate that the embryo itself controls much of embryogenesis. For example, most mutations that affect plant embryogenesis are associated with the genome of the embryo and not the mother plant. This is in contrast to embryo development in many animals where maternal factors have a significant impact on embryo development.

Still, a healthy parent plant is necessary for the successful development of the seed. If the maternal plant comes under environmental stress (such as from excessive heat, water stress, or nutrient deprivation) the resulting seed may not be viable.

INQUIRY LAUNCH

How could you test for the effect of environmental stress on the development of Fast Plant embryos?

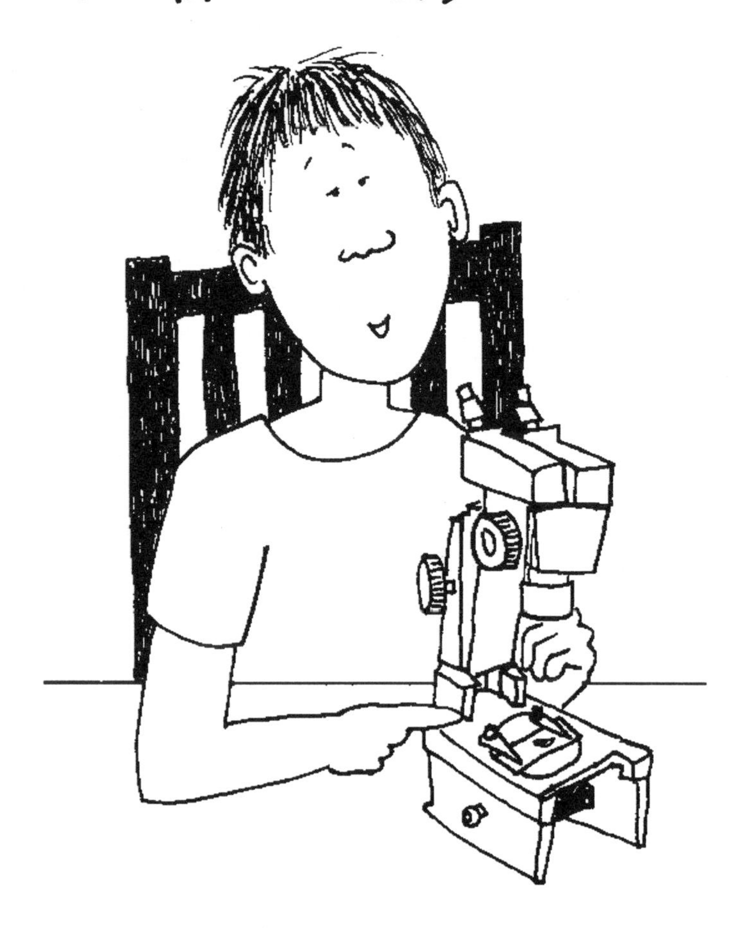

EMBRYO DEVELOPMENT in FAST PLANTS

BACKGROUND, continued

Stages of Embryo Development
Following fertilization in a Fast Plant, the 2n zygote undergoes several mitotic divisions. The first few divisions produce a *suspensor* (a strand of eight cells), which is attached to the embryo. The suspensor orients the developing embryo within the ovule and is thought to serve as an "umbilical cord," as it passes nutrients from the parent plant and from the endosperm to the embryo. The basal cell of the suspensor anchors the developing embryo and orients the embryonic root tip near the *micropyle* (the hole in the integuments where the pollen tube entered). At the tip of the suspensor, repeated cell divisions give rise to the very young *globular* embryo.

Immersed in the nutrient-rich endosperm, the Fast Plant embryo develops rapidly. By **Day 6 to 7**, the globular embryo becomes flattened and symmetrical. It has two lobes that will become the cotyledons, and root and shoot meristems. This is the *heart* stage.

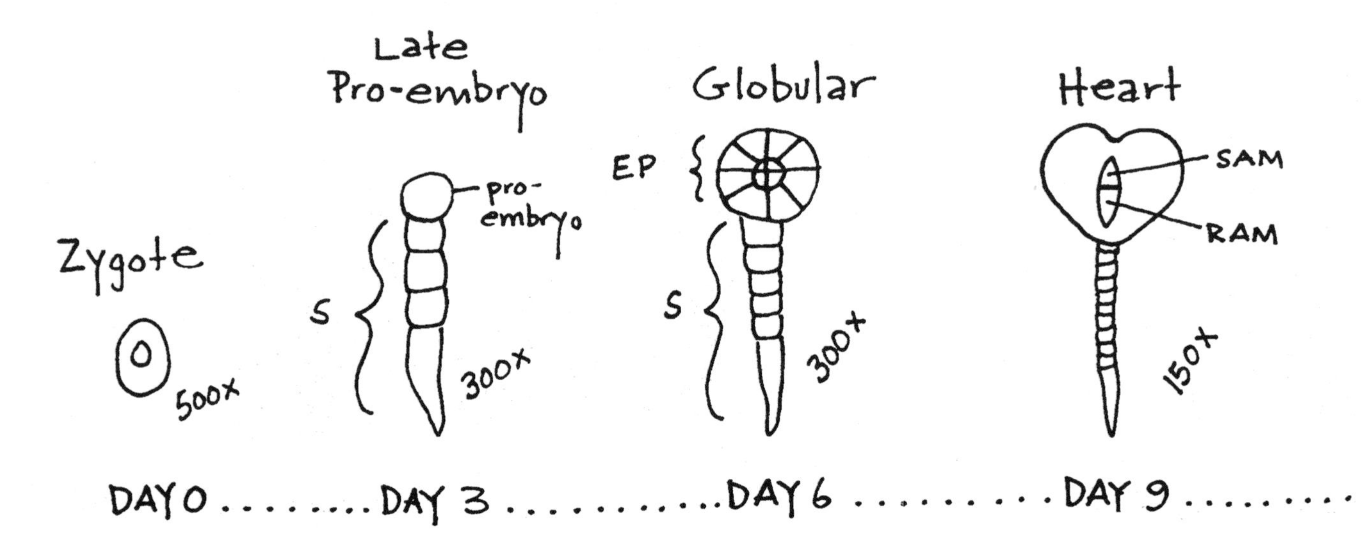

As its development continues, the embryo elongates into the *torpedo* stage. At this stage, the embryo produces chlorophyll and becomes green. Elongation of the embryonic hypocotyl separates the *root apical meristem* from the *shoot apical meristem*, which is hidden between the embryonic cotyledons. As the embryo enlarges, it consumes space formerly occupied by the endosperm. To package the enlarging embryo, the cotyledons fold around the hypocotyl, now curved within the ovule; this is the *walking stick* stage. As maturation proceeds in the enlarged folded embryo, the starch reserves within the embryonic cotyledons are converted to lipids as the final form of energy storage for seed germination in the future. By **Day 20**, the integuments harden and become the seed coat, and the embryo within desiccates to become a seed.

A SMALL BALL with MANY GENES

Food for Thought...

About 15,000 genes are expressed in a developing embryo. Many of these genes are only active in certain cells at specific developmental time points. Even at the globular stage of embryogenesis, differentiation of embryonic gene expression is present: distinct sets of genes are expressed in particular cellular regions.

Auxins (plant hormones that are involved in many aspects of plant growth and development) likely play a role in the control of embryo development and corresponding gene expression. They occur at greatest levels during the globular stage of embryogenesis and are subsequently located in a gradient from the embryonic shoot meristem down to the root tip.

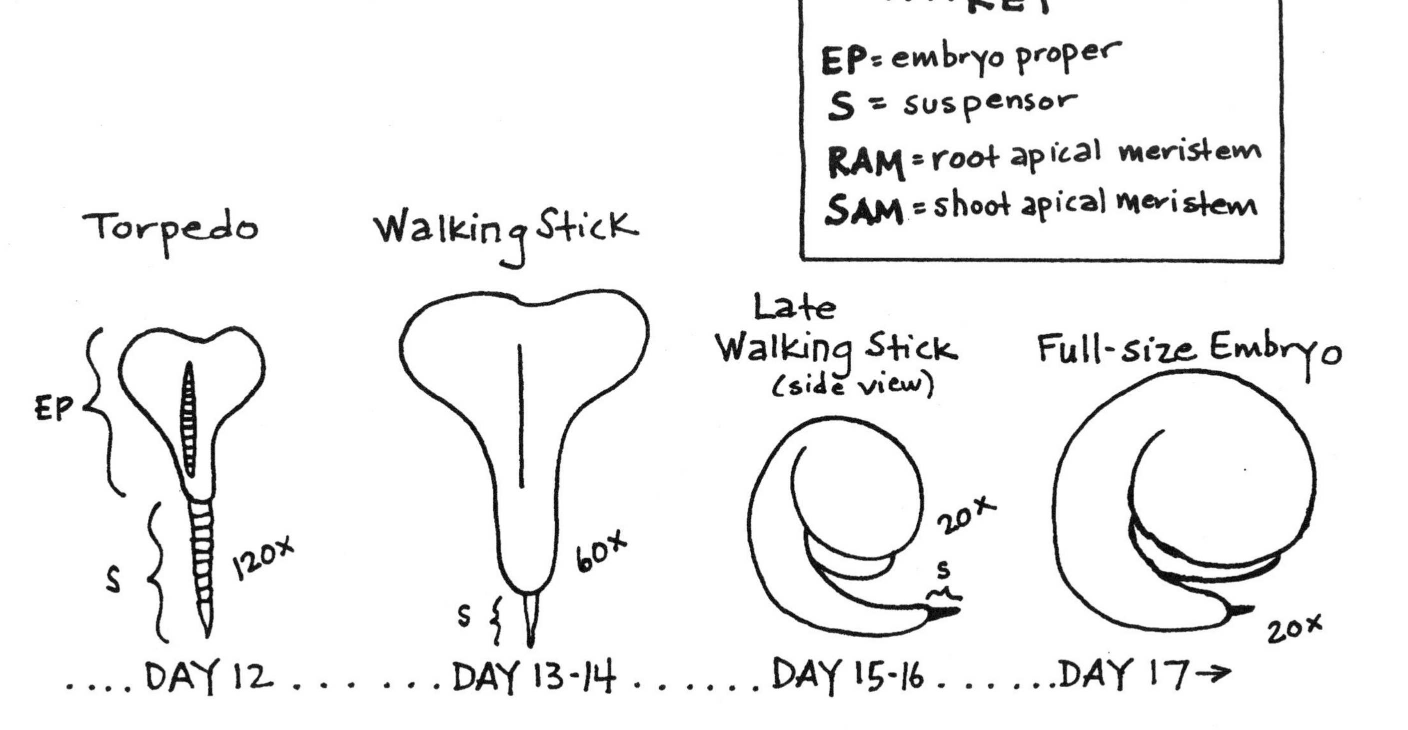

ACTIVITY

Activity 1
Embryogenesis—What, When and Where
Start this activity with two 14- to 16-day-old Fast Plants that have just been pollinated. (See Chapter 4, page 77, for more information if necessary.)

Procedure

1. On **Day 1** (0 *dap* or days after pollination), place a piece of tape on each wickpot. Label the pot with your name and number the plants.

2. Note that flowers on the plants are produced and open in a sequence spiraling up the plant stem. Starting with the oldest, lowest flower on the main stalk, number the flowers as 1, 2, 3, 4, and 5 with number 1 being the oldest flower.

   With a sharp pair of fine scissors, terminate flowering by snipping off all remaining flowers above flower number 5, leaving only the five open flowers that have been numbered. Snip off all side shoots. On subsequent days, you may need to terminate further flowering by snipping new buds, shoot apices, and side shoots as they appear.

3. Note the position of the stigmas relative to the tall anthers. Is the stigma below (–1), equal to (0), or above (+1) the tall anthers?
   ↪ Record this information for each flower of both plants on the Stigma Position and Pistil Length data sheet (page 104).

4. On **Day 3** (3 *dap*) notice that the flower parts that were important in pollination have withered and fallen

from the plant. Carefully measure the length of each expanding pistil.

↪ Record the pistil length to the nearest millimeter on the Stigma Position and Pistil Length data sheet (page 104). Remember the highest number (5) is the most apical (top) flower.

5. By **Day 6** the pistil is beginning to look like a young pod, which is exactly what it is!
   ↪ Measure and record the pod (pistil) length.

6. On **Day 9** notice that the pods, in addition to elongating, have swollen around what appear to be developing seeds or ovules. Carefully hold the plants up to the light. Can you see the outlines of the developing ovules within the pods?

7. By **Day 9** you will observe the developing embryos within the ovules.
   ↪ Measure and record pod length.

8. From either of your two plants, choose an older pod in which several ovules are visible and carefully snip it off with a fine scissors. Place the pod on a Scale Strip, aligning it longitudinally on the scale. (See Appendix 4, page 135, for information on Scale Strips.) Holding one end of the pod, use a sharp blade or needle to cut along one seam of the pod where the two carpels are fused. Pry open the pod to reveal the ovules aligned within the carpel; each ovule is attached to the vascular strands by its *funiculus*. You will also see a thin paperlike *septum* separating the carpels.

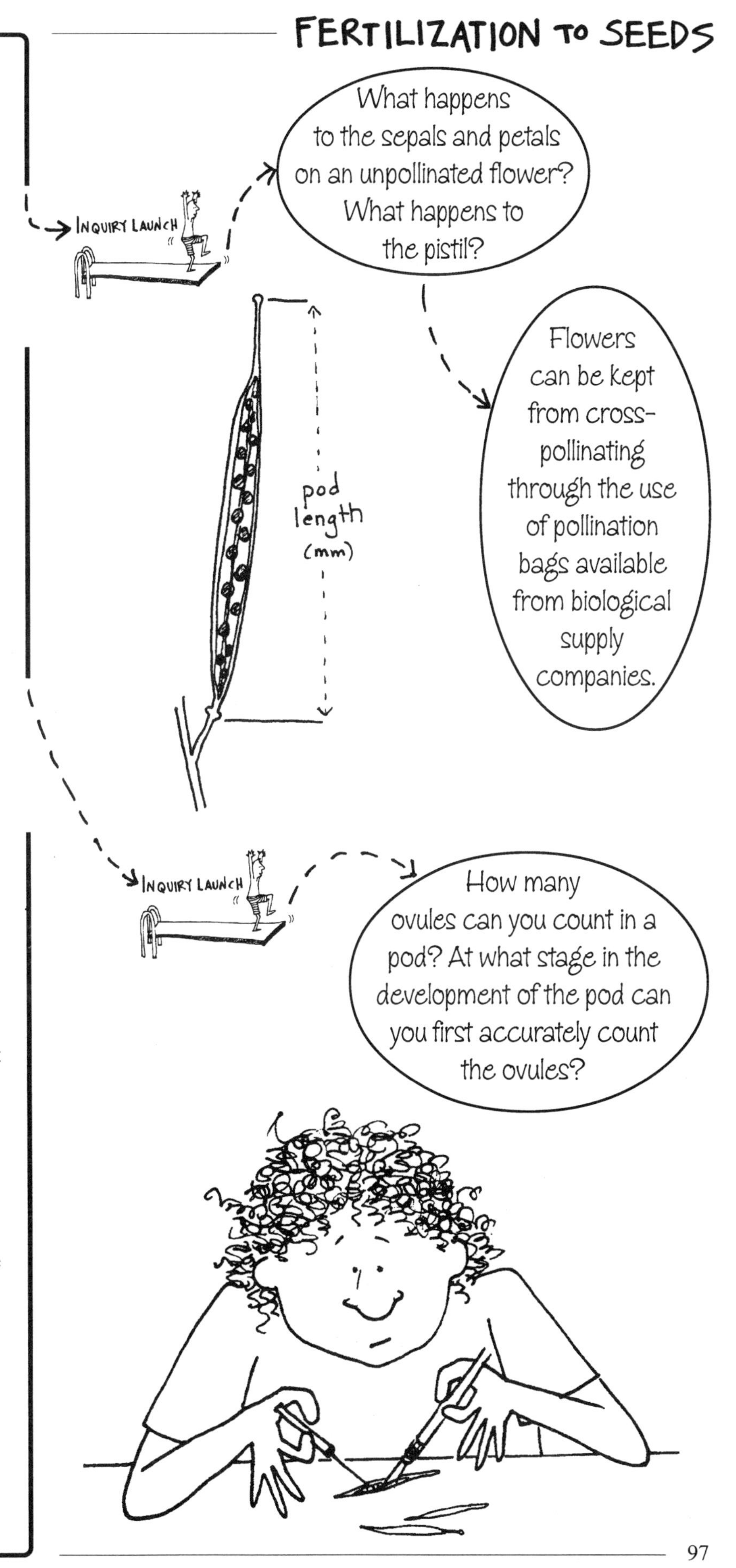

NOT JUST FOOD FOR THOUGHT

Food for Thought...

In Fast Plants and many other plant species, the food reserves of the endosperm are incorporated into the cotyledons during embryogenesis. These fleshy cotyledons in turn nourish the embryo as it grows during subsequent seed germination. In cereal crops (such as wheat, rice, millet, barley, oats, and corn), the endosperm is not used in embryogenesis by the enlarging embryo. Instead, it develops into a major portion of the seed as a starchy energy source for the germinating seedling.

Researchers have found that the quantity of energy reserves present in the cotyledons or endosperm of seeds not only nourishes the embryo during germination but is also a principal determinate of early seedling size. Partial removal of these energy sources causes a reduction in this size.

ACTIVITY, continued

9. With a dissecting needle or forceps, detach two or three ovules from the pod and place them in a cluster on the sticky tape of the Scale Strip. Put the pod aside on wet paper toweling to keep it fresh.

10. With a pipette, place a small drop of water over the ovules on the Scale Strip.

11. Place the Scale Strip under a dissecting microscope.

12. With two needles, hold and squeeze the first ovule, cutting into it with one needle. A tiny green object (the embryo) should pop out along with some cloudy material (the endosperm).

13. Draw a scale bar on the Seed Dissection sketch sheet (page 105), representing the distance of 1 mm or some fraction (0.5, 0.25, 0.1) of the magnified millimeter scaling. Indicate the distance represented by the bar on the drawing.
↪ Observe and draw the embryo to scale.
↪ Calculate and record the length of the embryo. Record the magnification of your microscope.
?? Does the embryo look like any of the illustrated stages in the embryogenesis illustration?
↪ If you didn't succeed with the first one, try another ovule.

14. On **Day 12**, the plant's lower leaves may be starting to turn yellow or even wither and dry. This is the normal

succession in the life cycle of the plant, particularly if it is supporting a number of pods with developing seeds.

↪ Measure and record pod length.

↪ Dissect embryo as outlined in steps 7–13 above. *Note*: If an ovule from this developmental time-point is illuminated from below, you may be able to see (through a dissecting scope) the indistinct embryo within the ovule.

15. On **Day 21**, plants are now approaching maturity. Normally the Fast Plant embryo has fully developed by 18 to 20 *dap*, at which time the stem and pods of the aging parent plant begin to turn yellow along with the leaves, and the seed coats begin to turn brown. This is the time that water can be withheld from the plants to encourage seed ripening.

↪ Make and record a final pod length measurement.

16. Based on your data, calculate the average pod length for 3, 6, 9, 12, and 21 *dap*. Enter this on the Pistil Length class data sheet (page 106).

PLANT GERIATRICS

Food for Thought...

As a plant ages and matures, certain tissues and organs become obsolete. From a casual observation it may seem that they are simply discarded by the plant. But plants actually initiate a whole sequence of events (*senescence*) before disposing of these tissues and organs through abscission.

Senescence is an active developmental process. At the genetic level, a set of *senescence associated genes* (SAGs) are expressed and a corresponding array of proteins are synthesized and become active. In turn, these proteins are responsible for the recovery of valuable resources (including amino acids, sugars, nucleosides, and minerals) from the tissues that are about to be discarded.

For annuals such as Fast Plants, the process of senescence allows the plants to transfer accumulated resources from vegetative tissues to the synthesis and maturation of seeds. In this instance, senescence is genetically preprogrammed. In other cases, such as the falling of leaves from trees in autumn, the trigger is mainly environmental.

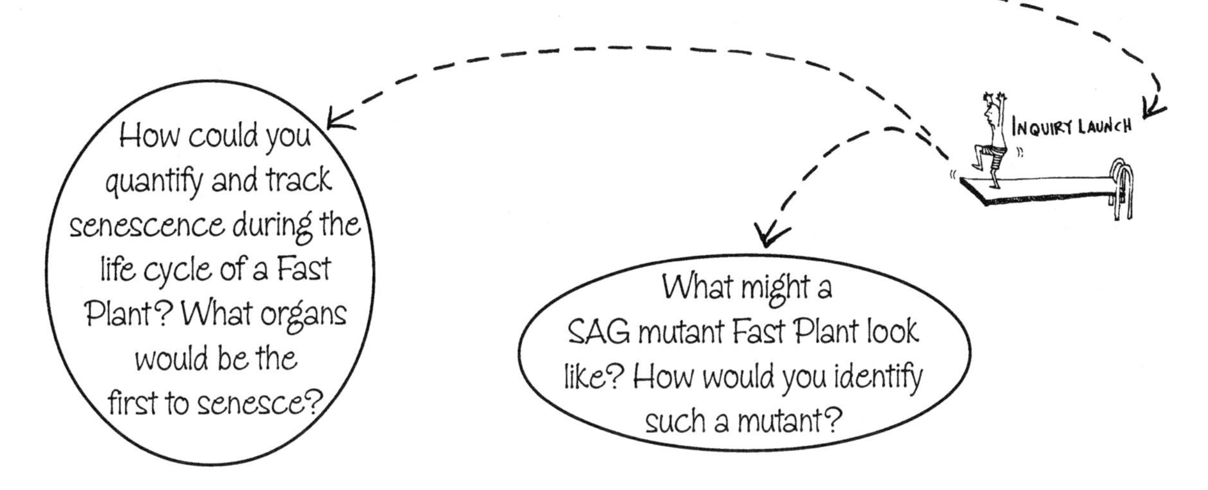

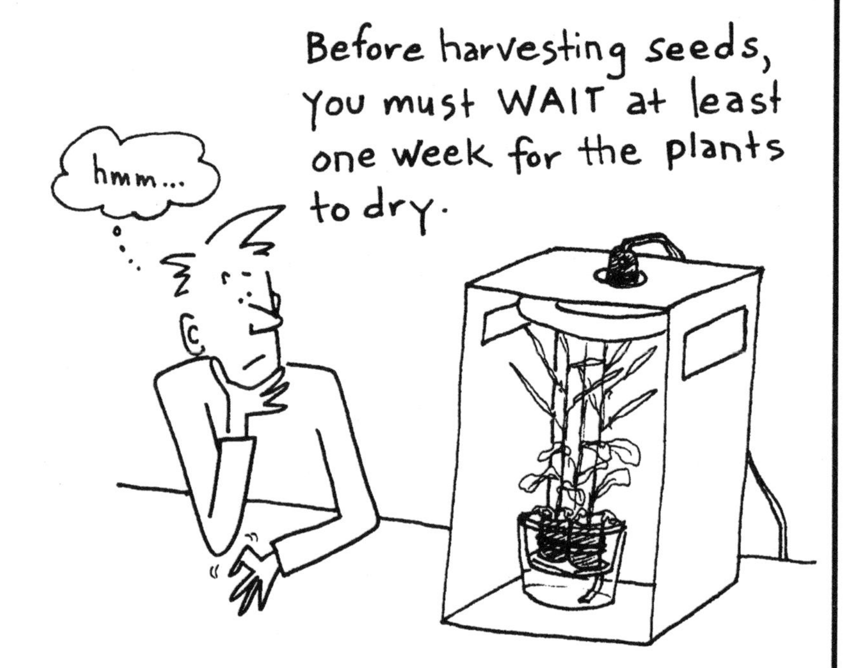

ACTIVITY, continued

↪ Calculate and record the average pod lengths for all plants in your class at 3, 6, 9, 12, and 21 *dap*.

↪ On the data sheet entitled, Pistil Length, Summary of Class Data (page 107), graph the average pod length at 3, 6, 9, 12, and 21 *dap*.

?? Is there a period in postfertilization development in which the increase in pod length is fastest?

?? Does the measured plant character of pod length on each day exhibit a normal distribution within the class population? Consider developing a class frequency histogram and statistical summary to answer this question. (See Appendix 2, page 127 for more information on normal distribution and developing class frequency histograms.)

17. Empty the water from the reservoir and remove the capillary mats. Keep the plants under the light bank and let them dry. You can use these plants and their seeds in many ways, as presented in the Epilogue (page 109).

Additional Reading

A. B. Bleeker & S. E. Patterson, "Last Exit: Senescence, Abscission, and Meristem Arrest in Arabidopsis," *The Plant Cell* **9**:1169–1179, 1997.

L. S. Dure, "Seed Formation," *Annual Review of Plant Physiology* **26**:259–278, 1975. A classic review on the subject of seed development with a focus on tissues associated with the ovule including the embryo and the endosperm.

R. G. Goldberg, G. de Paiva & R. Yadegari, "Plant Embryogenesis: Zygote to Seed," *Science* **266**:605–614, 1994. An excellent review of plant embryogenesis with an emphasis on genetic analysis and the plant species *Arabidopsis thaliana*.

L. D. Nooden & A. C. Leopold (editors), *Senescence and Aging in Plants*, Academic Press, San Diego, 1988.

M. A. L. West & J. J. Harada, "Embryogenesis in Higher Plants: An Overview," *The Plant Cell* **5**:1361–1369, 1993. An overview of plant embryogenesis with a conceptual focus.

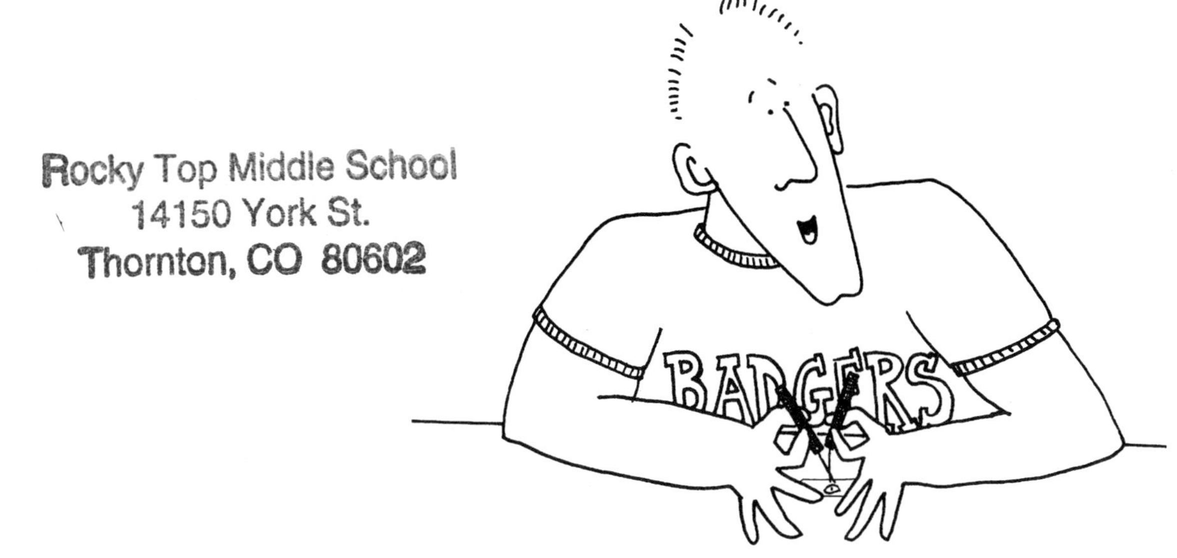

TEACHER PAGE

Overview

The activity in this chapter provides you and your students the opportunity to explore what happens between fertilization and seed harvest. In the activity, students have an opportunity to follow the results of student pollination. Pistil length is measured as an indicator of seed maturation and development. Embryo dissection provides the opportunity to observe embryogenesis and endosperm development.

Objectives

By participating in this unit students will:

- make observations and accurate measurements of pistil enlargement;
- dissect embryos from ovules in developing pods;
- make accurate descriptive observations of specimens under the microscope, draw carefully "to scale," and record and analyze data obtained from the drawings;
- understand the unique nature of fertilization in higher plants in which embryo and endosperm are interdependent specialized tissues within the ovule that function in normal seed development;
- understand that, following fertilization, a complex sequence of interdependent developmental events occurs that results in the production of viable seed for a new generation, thus continuing the life cycle;
- understand the interdependent relationship of developing maternal tissue and fertilized ovules; and
- learn that embryogenesis is a continuum of development from a very small spherical group of cells to a complex multidimensional, multicellular organism.

Time Required

Activity

Embryogenesis—What, When and Where

Seeds must be planted approximately 14–16 days* before activity (see calendar).

- **Day 0** (0 *dap*): 20 minutes to label, terminate flowering, and record pistil position.
  - **Day 3** (3 *dap*): 10 minutes to record pistil length.
  - **Day 6**: 10 minutes to record pistil length.
  - **Day 9**: 50–60 minutes to record pistil length, dissect and sketch embryo.
  - **Day 12**: 50–60 minutes to record pistil length, dissect and sketch embryo.
  - **Day 21**: 10 minutes to record pistil length.

The above estimates reflect only the time necessary for the laboratory activities. Additional time is necessary for data analysis and discussion.

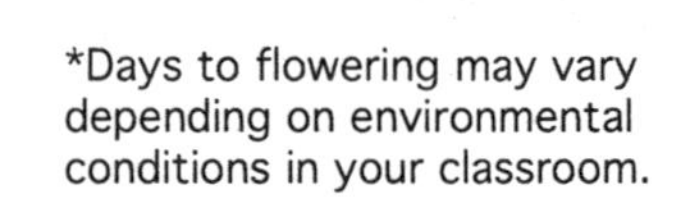
*Days to flowering may vary depending on environmental conditions in your classroom.

Materials

Each student will need:

Activity

- 2 Fast Plants, 14–16 *das*, pollinated
- dissecting microscope with 20 to 40X magnification
- 1 Scale Strip (Appendix 4, page 135)
- fine scissors
- fine-tipped forceps and fine dissecting needles
- clear double-stick tape
- glass microscope slide
- water and dropper
- sharp cutting blade

*See Appendix 1 (page 113) for growing instructions.

Classroom Management Tips

- The embryo dissection activity can also be performed using pods that have been harvested and "fixed" in an acetic alcohol fixative on specified days after pollination.
- To make acetic alcohol fixative, mix three parts 95% ethanol with one part glacial acetic acid.
- Fixed pods can be stored indefinitely for future use.

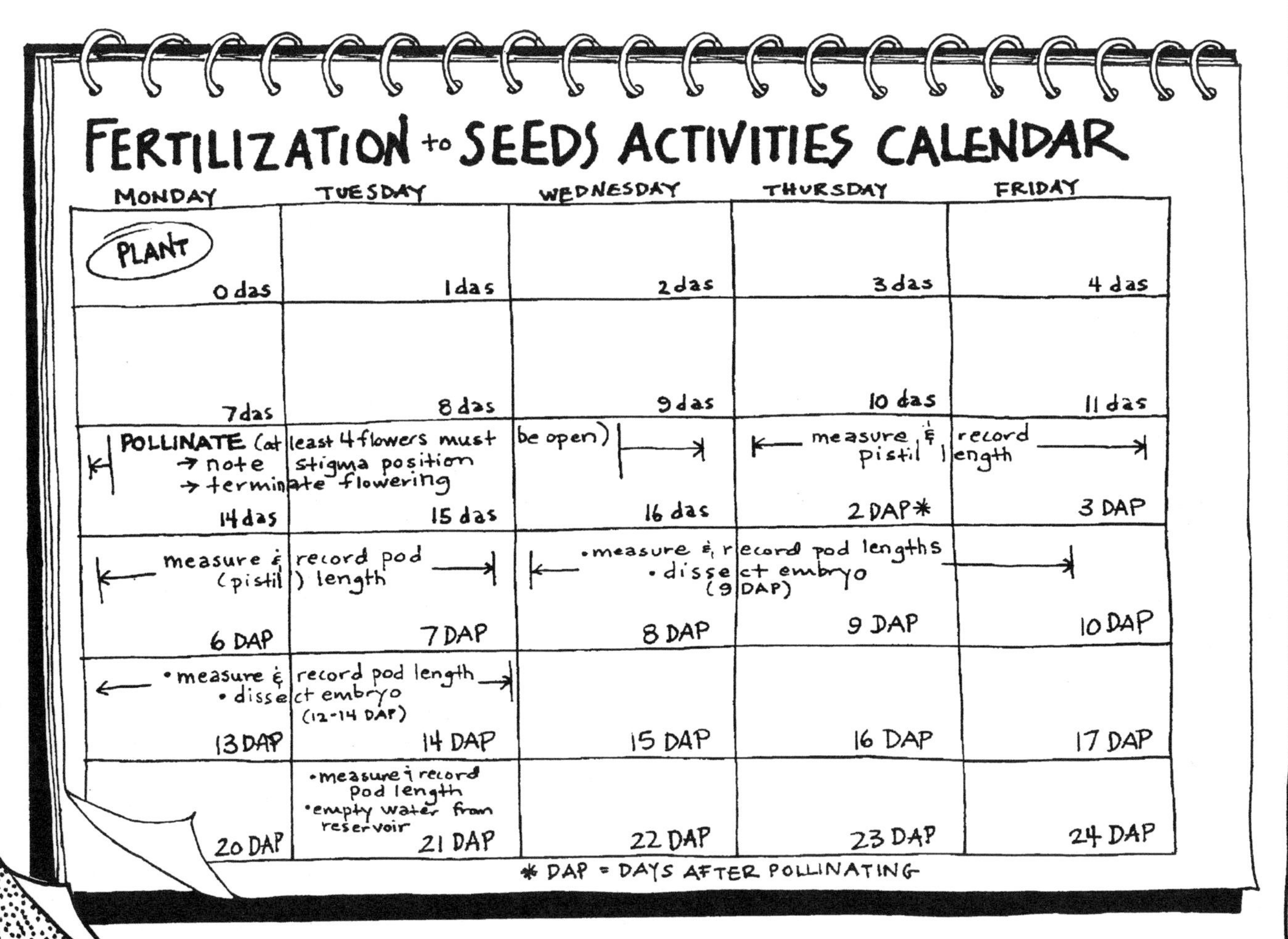

FERTILIZATION to SEEDS ACTIVITIES CALENDAR

MONDAY	TUESDAY	WEDNESDAY	THURSDAY	FRIDAY
PLANT 0 das	1 das	2 das	3 das	4 das
7 das	8 das	9 das	10 das	11 das
POLLINATE (at → note → termin 14 das	least 4 flowers must stigma position ate flowering 15 das	be open) 16 das	measure & pistil 2 DAP*	record length 3 DAP
measure & (pistil 6 DAP	record pod) length 7 DAP	• measure & r • disse (9 8 DAP	ecord pod lengths ct embryo DAP) 9 DAP	10 DAP
• measure & • disse 13 DAP	record pod length ct embryo (12-14 DAP) 14 DAP	15 DAP	16 DAP	17 DAP
20 DAP	• measure & record pod length • empty water from reservoir 21 DAP	22 DAP	23 DAP	24 DAP

* DAP = DAYS AFTER POLLINATING

Stigma Position and Pistil Length

Individual Data Sheet

How does pistil length change following pollination?

*dap**	Character	Plant #1 Flower number: 1	2	3	4	5	Plant #2 Flower number: 1	2	3	4	5
0	stigma position										
3	stigma/pod length										
6	stigma/pod length										
9	stigma/pod length										
12	stigma/pod length										
21	stigma/pod length										

* *dap* = days after pollinating; measurements should be taken at approximately 0, 3, 6, 9, 12, and 21 *dap,* but the actual measurement days may vary slightly due to the development of your plants.

Statistics on Pistil Length

Day	*dap*	Pistil Length n	$\overline{x}$	r	s

n = number of measurements, r = range (maximum minus minimum),
$\overline{x}$ = mean (average), s = standard deviation

Seed Dissection

Sketch Sheet

What stages of embryogenesis can you find?

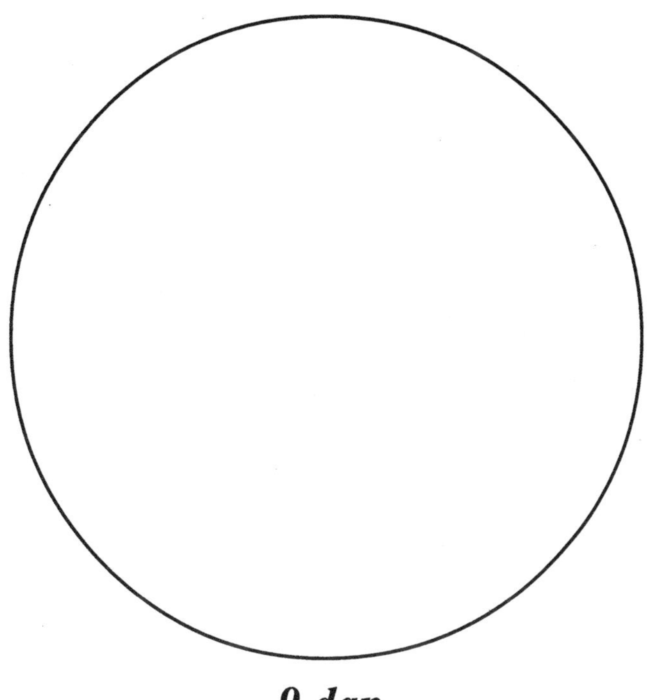

length of embryo __________
magnification of drawing __________

9 dap

Sketch embryo, include scale bar

length of embryo __________
magnification of drawing __________

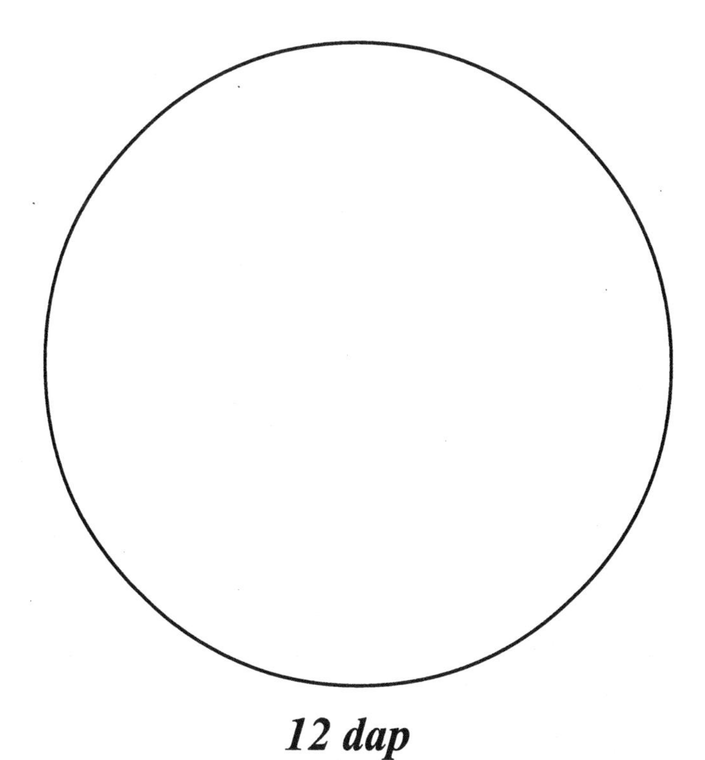

12 dap

Sketch embryo, include scale bar

Pistil Length
Class Data Sheet

Name	Average of all observations 3* *dap*	Average of all observations 6* *dap*	Average of all observations 9* *dap*	Average of all observations 12* *dap*	Average of all observations 21* *dap*
Class Average					

* measurements should be taken at approximately 3, 6, 9, 12, and 21 *dap* but the actual measurement days may vary slightly due to the development of the plant.

Pistil Length

Summary of Class Data

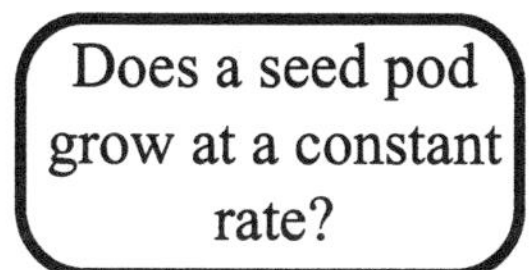

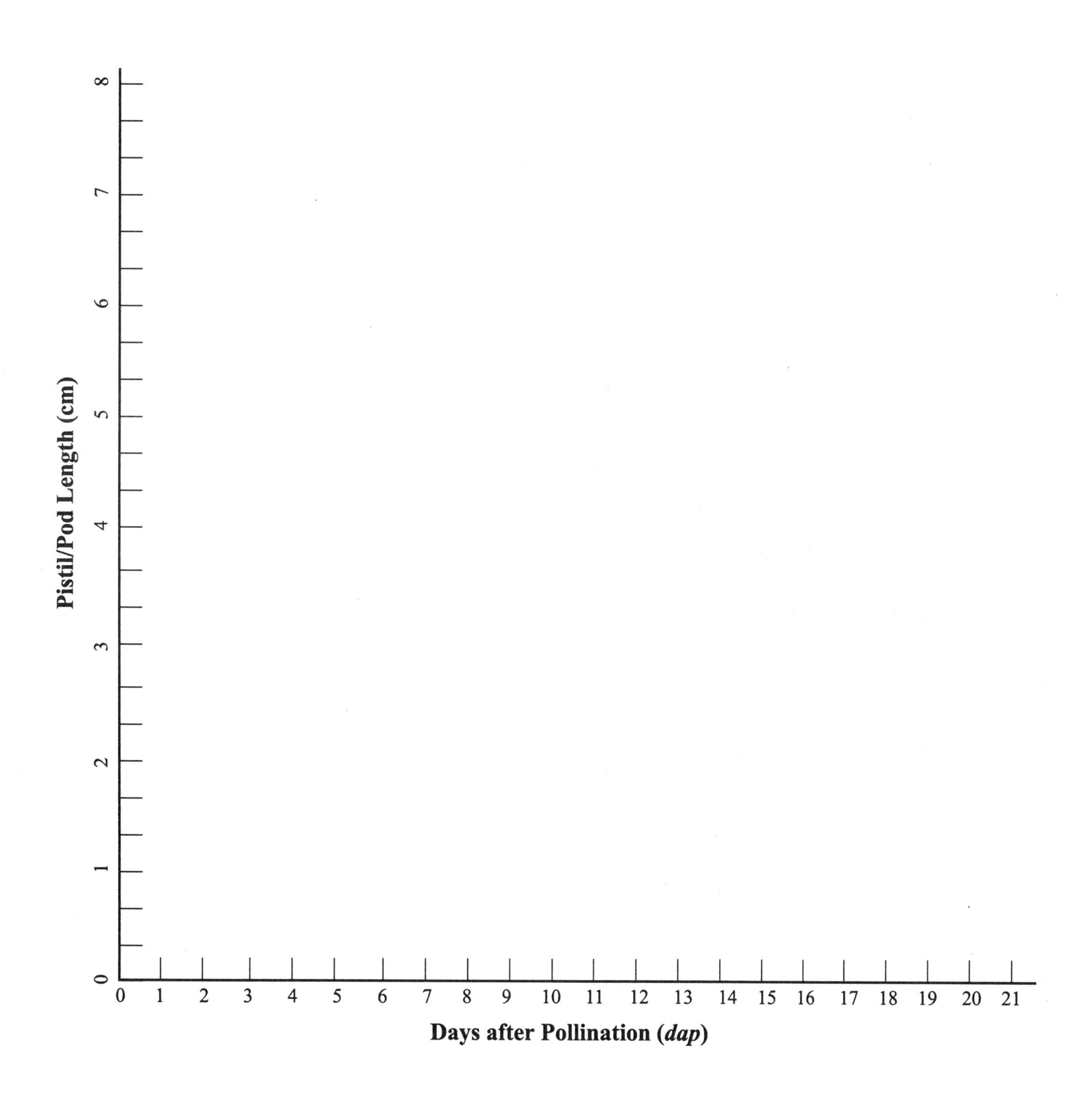

EPILOGUE

Follow the spiral and track variation

Understanding how variation is generated and maintained is central to an appreciation of biology and is an important theme in the study of genetics. A "family approach" to Fast Plant investigation provides a rich and understandable introduction to the complexities underlying biological variation.

After sowing, growing, and pollinating Fast Plants, the seeds grow and develop. Twenty days after the final pollination (approximately Day 38), water should be removed from the reservoirs. About a week later, the pods on your plants will be dry. Gently remove each dried pod and place them between separate pieces of folded, clear adhesive tape being careful not to lose any seeds. Record the plant and pod number on the tape.

stigma
style length (mm)
pod length (mm)
ovary (carpel) length (mm)
receptacle
flower stem

POD CHARACTERISTICS

The plant and the pods can be viewed as the "mother." Since each of the seeds is an offspring of a common maternal parent, they are siblings within a family.

?? Are the seeds from a given plant half-siblings or full-siblings? (Do they have the same father?) What are the genetic implications of this?

?? Are the siblings from a single pod more like each other than they are like the siblings from another pod on the same plant? Explain your answer from a genetic point of view.

Seeds can be stored in the tape strips for months if kept dry and cool (e.g., in a refrigerator). When you are ready, the pod can be thoroughly crushed within the tape. As the tape is peeled open, the seeds will stick to the adhesive. They are ready to be planted.

With a new generation in hand, you are ready to spiral through the Fast Plants life cycle again! But this time you have an invaluable tool. Not only do you have seeds which allow you to investigate the stages of the Fast Plant life cycle, but you have a population with known genetic relationships that will allow you to go much further than before.

Throughout all of the life cycle, plant populations display variation. Explore the variation between and among individuals, families, and populations of Fast Plants. In what ways is variation due to genetic factors? In what ways is it due to environmental factors?

But where do you want to go? What part of the life cycle do you wish to investigate? What questions do you want to ask?

What percentage of Fast Plant seeds germinate under ideal conditions?

How does a germinating Fast Plant seedling respond to gravity?

How does a Fast Plant respond to light?

How does light color affect germination?

Does the root or shoot grow faster in a Fast Plant during germination?

GERMINATION

VARIATION: Genetic or Environmental? or both?

How rapidly does the pistil of a pollinated Fast Plant grow?

How many seeds does a Fast Plant produce?

How fast does a pollen tube grow?

When do a Fast Plant's petals first fall off?

When do the first organs on a Fast Plant senesce?

FERTILIZATION TO SEEDS

GROWTH AND DEVELOPMENT

What happens if the apical meristem is removed?

How many hairs are on the margin of a Fast Plant leaf?

Does a plant from a deeply sown seed emerge later?

How tall is a Fast Plant at different life stages?

What is the sugar concentration of the nectar?

What is the effect of temperature on time to first flower for a Fast Plant?

How much nectar is produced?

When does the first flower open?

FLOWERING

How does humidity affect pollen viability?

How old can a Fast Plant flower be and still be successfully pollinated?

How quickly does a pollen tube germinate ?

POLLINATION

GROWING & MAINTAINING FAST PLANTS

INTRODUCTION

To optimize the Fast Plants experience and to prevent any disappointment, please read this section thoroughly prior to conducting any experiment. Fast Plants are easy to grow and will perform as expected if certain guidelines are followed throughout the life cycle.

The following section provides information on:

- time requirements,
- setting up a proper lighting system and a growing system,
- preparing the nutrient delivery system (fertilizer) and planting medium,
- planting Fast Plants,
- Fast Plants care,
- harvesting Fast Plants,
- trouble shooting and tips, and
- material sources and Fast Plants seed stocks.

To keep within the philosophy of the Wisconsin Fast Plants Program, you should feel free to develop your own growing systems using low-cost, readily accessible materials. For more information on growing and maintaining Fast Plants, visit the WFP website at **www.fastplants.org**.

Time Requirements

How long will it take to grow Fast Plants? The entire life cycle takes 35–45 days, from planting to harvesting seed. From day to day, the amount of time required varies, depending on the task and activities you plan to do with your students (see specific chapters and activities for details). For example, on the days when students are just observing their plants, checking the water level in the reservoirs, or making notes in their journals, only a few minutes are required. Activities that may require more time include planting, data collection, pollinating, and seed harvesting. After the flowers are pollinated, the plants require little care (except for watering) until the day the seeds are harvested.

Lighting and Growing Systems

Perhaps the most critical component for growing Fast Plants is 24 hours of intense fluorescent lighting. **Fast Plants will NOT be successful when grown with the light available on a windowsill, in the classroom with normal overhead lighting, or in a greenhouse without supplemental lighting.** This is very important in order to grow strong, healthy plants and complete the growth cycle in 35–45 days.

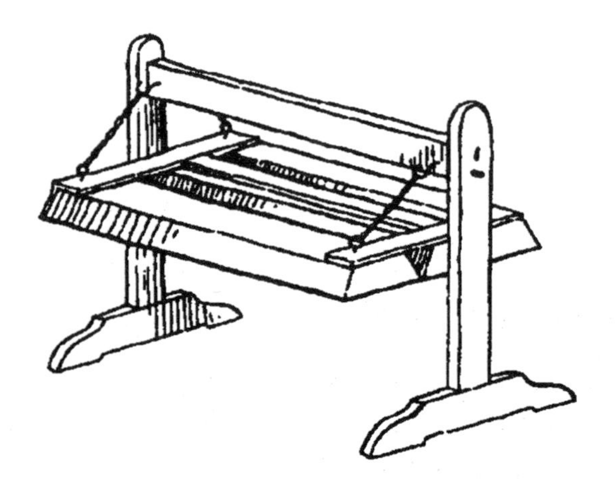

Fast Plants can be grown under commercially available fluorescent light banks, or you can construct your own based on the space available in your classroom. We have found that the most effective lighting systems are 6-bulb, 4-foot banks, purchased from home supply stores either as a 6-bulb unit or put together from three 2-bulb shop lights. Cool white, 40 watt bulbs or high-efficiency T-8 bulbs are recommended, depending on the fixture you choose. We include instructions here for the Plant Light House as an option to the larger banks. For other Fast Plants lighting options, check the Fast Plants website at **www.fastplants.org**

Plants should be kept with their growing tips within 5–10 centimeters of the light bulbs. We recommend keeping the lights at a constant height and using books or other materials to raise and lower the plants as they grow (i.e., moving the plants up and down is much simpler and less likely to cause costly accidents than raising and lowering light banks).

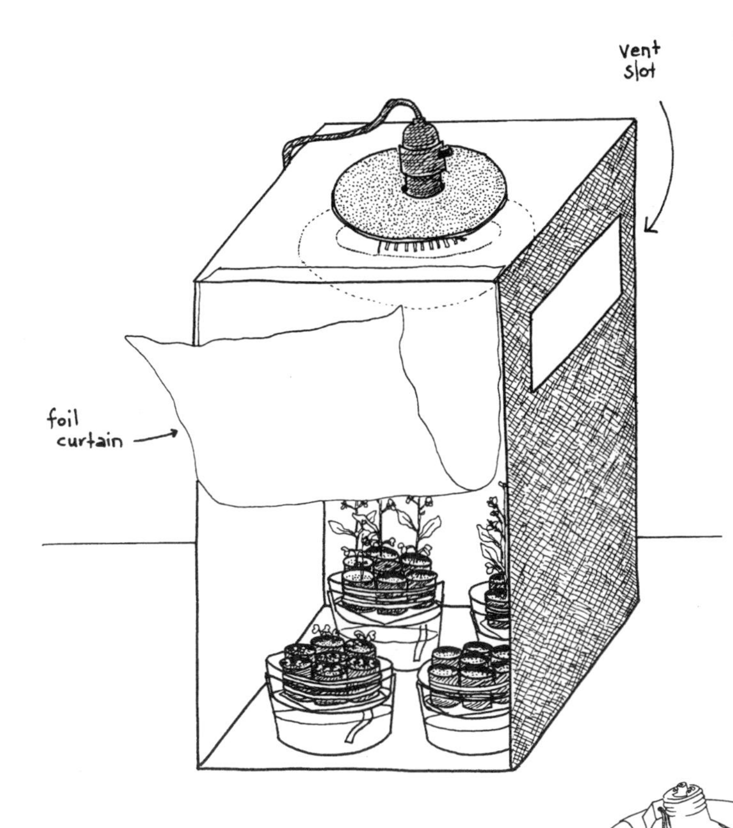

Plant Light House Construction

Materials

For each Plant Light House, you will need:

- one empty copy paper (8.5 x 11 inch) box
- aluminum foil roll, 12 inches wide
- glue stick or white glue
- clear tape
- cutting blade
- scissors
- soft plastic container lid or plastic plate
- 30-watt fluorescent light or 39-watt energy saver fluorescent light
- electric cord with socket

30-39 watt

socket plug

socket cord

Construction

1. Prior to construction, inspect box and reglue any weakly glued flaps.
2. Apply a thin layer of glue to inner surfaces of the box and attach aluminum foil to cover the <u>entire</u> inner surface.
3. Cut 3 vent slots (4 x 14 cm) in the top of sides and back, as shown.
4. Cut a hole (2.5 cm diameter) in the center of container lid and trim off the rim to make a flat disc with a hole.
5. Insert the light fixture through hole from inside the box.
6. Place disk over light fixture from the top and attach the plug and socket.
7. Tape aluminum foil curtain to top front edge of box.
8. Strengthen edges and center of curtain with clear tape.

Potting Mix Preparation

Into a large container, pour a sufficient amount of peatlite potting mix (see page 125). Mix well with your hands until the planting medium is slightly moist and loose. Add small amounts of water if necessary. The potting mix should feel moist, be loose and fluffy, and should not clump together or drip water when squeezed in your hands.

Water Bottle Construction

Materials

- one soda bottle (0.5 to 1.0 liter)
- small nail
- pliers
- propane torch or candle
- drill and 1/16-inch drill bit (optional)

Construction

1. While carefully holding it with the pliers, heat the nail in a flame. Use the hot nail to melt a hole in the center of the bottle cap (1–2 mm diameter). Alternatively, drill a 1/16-inch hole in the center of the bottle cap.

2. Fill bottle with water and replace the cap. Turn the bottle upside-down and squeeze!

Nutrient Solution Preparation

If you choose to use liquid fertilizer solution in your reservoirs rather than the slow-release pellets in your wickpots, use the following instructions to mix the solution.

Materials

- 2-liter soda bottle or other large bottle with cap
- permanent marking pen
- kitchen funnel
- measuring spoons
- Peters Professional™ or Miracle Gro™ fertilizer (see page 125)

Preparation

1. Remove the bottle labels by loosening the glue. Fill the bottles with hot tap water or use hot air from hair dryer to soften the glue, and then peel the labels off.

2. Make a stock solution by mixing 1/4 teaspoon of the fertilizer crystals per liter of water. This is the 1X strength solution; you can make other solutions that have a higher or lower concentration if you wish to investigate the effects of fertilizer on the growth of Fast Plants.

3. Fill your reservoirs with the solution. The plants will take up the nutrients via capillary action.

4. Check your reservoirs and replenish the nutrient solution regularly.

Store your nutrient solutions away from any light to reduce algae growth.

Growing System Construction

The instructions presented here are for the quad wickpot growing system provided in kits from Carolina Biological Supply Company. If you wish to use an alternate growing system, you can visit the Fast Plants website for instructions and suggestions.

Materials

- styrofoam quads
- quad wicks (four per quad)
- reservoir with lid
- capillary mat
- water

Preparation

1. Drop a diamond wick into each quad cell. Pull the wick from the bottom through the hole, so that the wick is halfway inside the quad. Wet each wick with water from your water bottle.

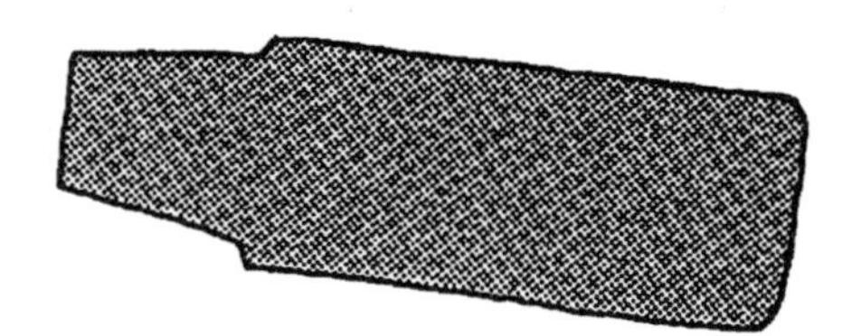

2. Thoroughly wet the capillary mat for the reservoir. There should be no visible air pockets. Place the mat on top of the reservoir lid.

3. Fill the reservoir either with water (if you are using the slow-release pellet fertilizer) or the nutrient solution. Snap the lid onto the reservoir with the tail of the capillary mat extending into the liquid.

Your growing system is ready!

Planting Fast Plants

Materials

- growing system (above)
- planting mix (page 116)
- slow-release fertilizer pellets OR fertilizer solution (Do not use both!)
- water bottle (page 116)
- straight edge or ruler
- Fast Plants seed (page 125)
- light bank or Plant Light House (page 115)

Procedure

1. Loosely fill wickpots halfway with prepared planting mix.

2. Drop three fertilizer pellets into each cell. Fill the cells with potting mix, covering the pellets.

3. Gently tap film can wickpots with fingers to settle planting mix.

4. Carefully scrape off all excess soil. **Do not press or compact the soil.**

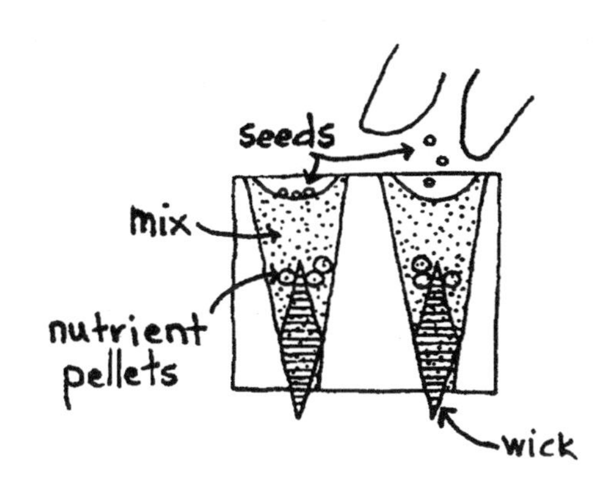

5. Water the planting mix until it drips from the wicks at the bottom of the quad (the planting mix will shrink down with watering).

6. Place 3–4 seeds on the surface of planting mix of each wickpot. Make sure the seeds are evenly spaced.

7. Sprinkle just enough planting mix over the seeds to cover them. **Do not press or compact the soil.** Water the wickpots gently so that the seeds are not washed away.

8. Place the wickpots onto the wet capillary mat on the reservoir.

9. Place a label in each wickpot with the following information:
   - type of seed,
   - your initials, and
   - date of planting.

10. Place the growing system under your lights.

Fast Plants Care

Materials

- water or nutrient solution
- scissors or forceps
- 10" bamboo skewers (available at grocery stores)
- thread or twist-ties for tieing up plants
- pollination supplies (beesticks, etc.)

Every Day

Make sure your reservoirs are adequately filled with water or nutrient solution (if you are using nutrient solution rather than the slow-release pellets). **Be sure to fill the reservoirs fully before weekends or school holidays.** As the plants grow, they will use more of the water or nutrient solution each day. By Day 10 in the life cycle the plants may use a full reservoir every 2–3 days.

5–7 Days After Sowing

Thin to 1 plant per quad cell or 2 plants per film can by cutting off extra plants with scissors just above the surface of the planting medium.

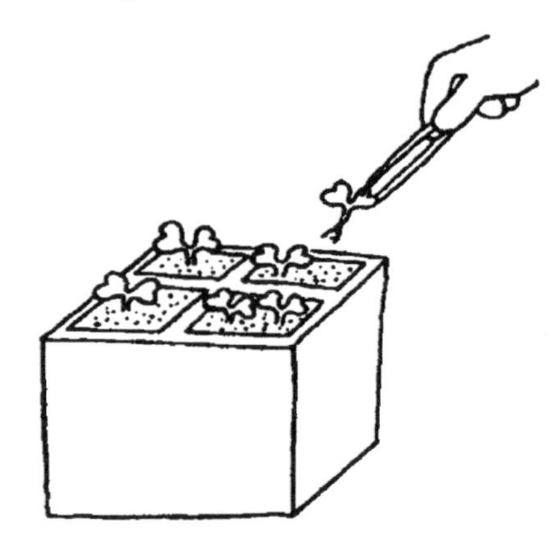

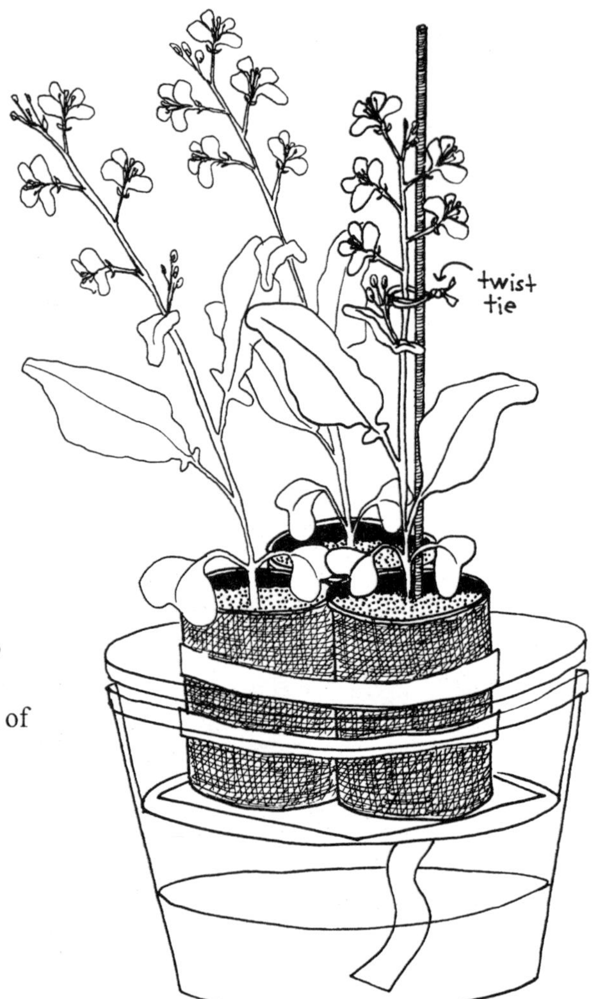

7–14 Days After Sowing

Some plants may need to be staked up with bamboo skewers to keep them from falling over. Place a skewer next to the plant stem, press it down through the planting medium to the bottom of the wickpot, and secure the plant to the skewer with ties.

15–18 Days After Sowing

Plants must be pollinated to produce seed (see Chapter 4, Pollination).

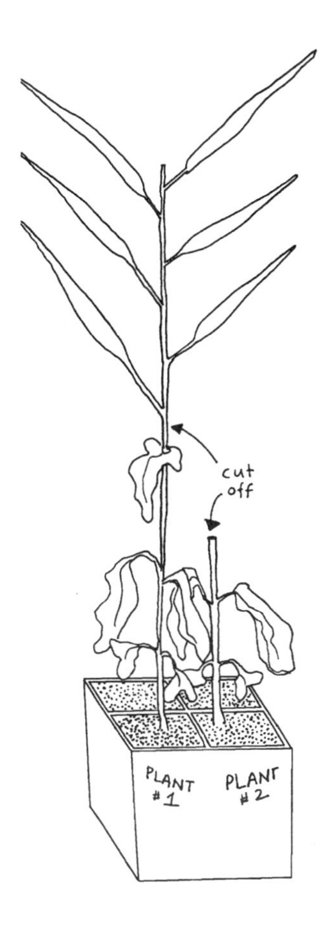

Harvesting Fast Plants

Materials

- scissors
- brown paper bags
- pencil
- shallow plastic tray or metal pan
- seed envelopes (coin envelopes)
- stapler
- zipper-type plastic sandwich bags
- indicating silica gel drying compound (see page 125)

Procedure

1. Approximately 20 days after last pollination, when the ends of the pods are changing from green to brown, remove the wickpots from reservoir or empty the reservoir and leave the wickpots there to dry.

2. Let plants dry for 7 days until the pods are crisp and brown.

3. Cut the plants off and place them in a paper bag. Label the bag with the planting and pollination information.

4. If pods are not crisp, let them dry further in bag. Staple the bag shut.

Note: If you plan to start your next Fast Plants unit with pods rather than seeds, you may choose to keep the pods attached to the mother plants. If not, proceed through the next steps.

5. Break up the pods thoroughly by crushing them in the bag to release the seeds.

6. Pour seed and chaff into the shallow tray.

7. Pick out the large pieces of stems and leaves, and the remaining pod pieces.

8. Gently blow on the remaining mixture. The chaff should blow away, leaving clean seeds. You may wish to do this outside.

9. Place your clean, dried seeds into a labeled envelope.

10. Store your seed envelopes in a zipper-type sandwich bags in a refrigerator. For optimal long-term (12-month) storage, add silica gel in the bag to remove any remaining moisture. Seeds stored under these conditions will remain viable for many years.

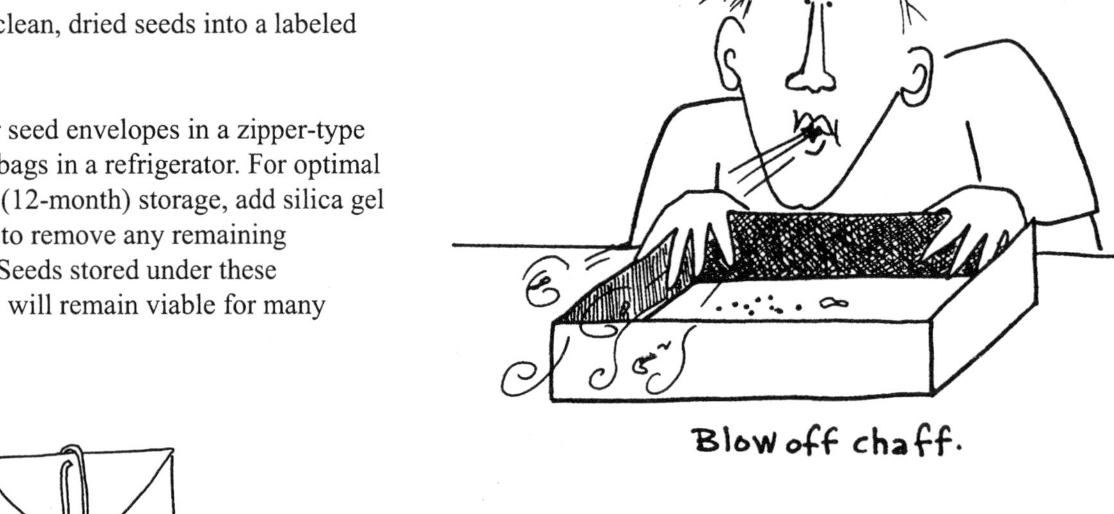
Blow off chaff.

Label envelope of clean seeds.

Cleaning Up and Reusing Materials

All growing systems, wickpots, capillary mat, and water bottles can be reused. Do not reuse planting medium.

To clean your planting materials:
1. Rinse off as much of the residual planting medium as you can.
2. Soak all materials in a 10% bleach solution for at least 20 minutes.
3. Rinse well and let air dry.

Troubleshooting and Tips

The most commonly reported problems are listed below, along with their possible causes.

Poor germination (no seedling emergence)

- Seeds planted too deeply in wickpot.
- Planting mix compacted or too wet when filling wickpots.
- Wickpot not watered sufficiently after planting.
- Seeds washed out of wickpot.
- Student planted fertilizer pellets instead of seeds (it happens!).
- Room temperature below 15.5°C (60°F).

If seedlings do not appear by Day 4, start over.

Plants growing slowly

- Lower temperature in school than normal on weekends and holidays.
- Insufficient light. Plants not grown under recommended lighting conditions.
- Plants growing at lower temperature due to location near window in winter.
- Poor capillary action between plants and reservoir.

Plants look spindly

- Lights too far away from plants. Ideally, the plant growing tips should be 5–10 cm from lights.
- Fertilizer not delivered properly.
- Poor capillary action between plants and nutrient reservoir.

Plants wilt

- If the worst happens (you forgot to fill the reservoir) and the plants are wilting but not yet crispy, you may be able to save them. Water the wickpots gently from above for several minutes. Be sure that the wicks are dripping and the planting mix is moist. If the planting mix has completely dried, it may be difficult to remoisten. Rewet the capillary mat and tail and place the wickpots back on the mat.

Plants die

- Wicks not placed correctly in bottom of wickpots.
- Tail on the capillary mat not in the reservoir water.
- Capillary mat not wet thoroughly. All air pockets not removed when watering system was set up.
- Capillary mat clogged and not wicking water. Wash mat in 10% bleach solution, rinse thoroughly.
- Wickpot not completely in contact with the capillary mat.
- Water in reservoir ran out over weekend (always check water on Fridays!).
- Plant damaged during thinning (handle gently).
- Plant damaged during movement (as plants grow taller, stake and secure them with twist ties or thread).

No seed production

- Pollination not adequately performed.
- Too much heat in the classroom during pollination period. When temperatures are above 29°C (85°F), Fast Plants will lose the capability of producing pollen.

MATERIAL SOURCES

The following materials are mentioned in this manual for growing Fast Plants or completing the activities. You should feel free to modify materials and create your own systems as you wish.

Fast Plants seed—Available from Carolina Biological Supply Company (1-800-334-5551). See page 126 for seed stock information.

Quads and reservoirs—Available from Carolina Biological Supply Company (1-800-334-5551).

Fluorescent light and socket—Available from local hardware stores, home supply stores, or discount stores. Lights of America 30-watt or the General Electric 39-watt are recommended; 23-watt compact fluorescent bulbs also work well.

Film Cans—35-mm film cans are available from your local camera store or film processing outlets. The cans are usually discarded or recycled, so ask to have them saved for you.

Capillary mat—The recommended capillary mat, known as WaterMat™, is available from Carolina Biological Supply Company (1-800-334-5551). Alternative material is interfacing from a local fabric store. We recommend washing this material several times prior to using to break down the fire retardant on the material. Test these materials for capillary wicking capacity prior to using.

Fertilizer—Osmocote pellets (20-20-20 N-P-K) are available from Carolina Biological Supply Company (1-800-334-5551) or from your local garden supply store. Peters Professional™ or Miracle Gro™ fertilizer is available from gardening supply stores. Other fertilizers may work just as well, however, they have not been tested for the activities in this manual.

Potting mix (peatlite)—Commercial brands such as Jiffy Mix™ or Scotts Redi-Earth™ are available from your local garden supply store. Peatlite is a mixture of 50% peat moss to 50% vermiculite. Premixed planting medium is available from Carolina Biological Supply Company (1-800-334-5551).

Bees—Honeybees can be obtained from local beekeepers, or are commercially available from Carolina Biological Supply Company (1-800-334-5551).

Indicating Silica Gel Drying Compound—Type III indicating silica gel changes color from blue to pink when stored as above 20% relative humidity. Available from most chemical supply companies including Sigma Chemical (1-800-835-3010) and Aldrich Chemical Company (1-800-558-9160).

Fast Plants Seed Stocks

The following seed stocks* are available through Carolina Biological Supply Company (1-800-334-5551).

Seed Stock	Description	Main Uses	Genetic Designation
Standard (Improved Basic)	standard Fast Plant for form and performance; flowers in 13-17 days, variable expression of purple pigment in stems	life cycles, environmental science, botany, comparative morphology, physiology	Rbr. Rapid-cycling *Brassica rapa*
Purple Stem, Hairy (High Anthocyanin)	dominant expression of purple pigment in stems (also in petioles and bud tips); also selected for higher-than-usual number of hairs	genetic studies of purple pigment (anthocyanin) production and expression in crosses with Non-Purple Stem mutant, study of inheritance of leaf hairs	*ANL,* Hir (5-8)
Non-Purple Stem (Anthocyaninless)	recessive gene blocks the expression of purple, red, or pink pigment, also selected for few or no hairs	genetics (see Purple Stem, Hairy)	*anl,* Hir (0-1)
Yellow-Green Leaf	recessive, leaves yellow-green, purple pigment in stems	genetics, photosynthesis	*ygr*
Non-Purple Stem, Yellow-Green Leaf	double recessive mutant, with yellow-green leaves and no purple pigment expressed in plant	genetics, photosynthesis	*ygr, anl*
Rosette-Dwarf	recessive gene, gibberellin-deficient mutant, compact due to shortened internodes; attains normal height after application of 100 ppm gibberellic acid; flowering in 18 days	genetics, physiology	*ros*
Tall Plant (Elongated Internode)	recessive, abnormally tall due to elongated internodes, gibberellin-overproducer; may require staking	genetics, physiology	*ein*
Petite	recessive; plants reduced to a height of 5-15 cm; not a gibberellin-deficient mutant	genetics, physiology	dwf_1
AstroPlant	recessive; variant of Petite stock, shorter and more uniform in height (10 cm)	genetics, research in space	dwf_1
Variegated (Somatic Variegated)	maternal inheritance (chloroplasts); irregular patches of green white on green leaves	maternal (cytoplasmic) inheritance, photosynthesis	Var

*For a complete listing of all the Fast Plants seeds stocks and detailed descriptions, refer to the Fast Plants website at **www.fastplants.org.** Several seed stocks mentioned in this manual are not available through Carolina Biological Supply Company. For ordering information for the *apetalous* and *male sterile* seed stocks, contact the Wisconsin Fast Plants Program through the Fast Plants website.

APPENDIX 2

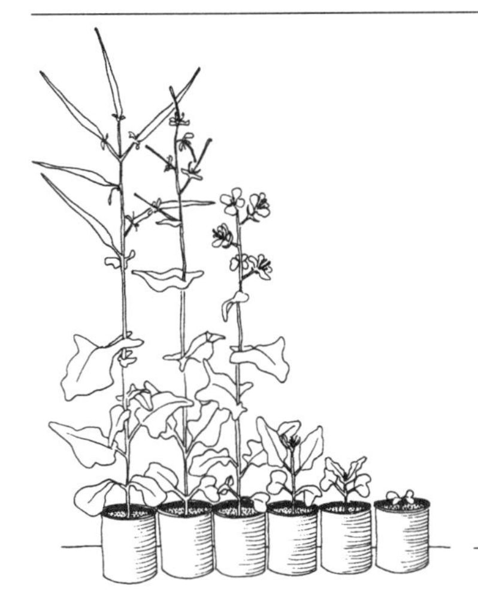

VARIATION

Collecting, organizing, and interpreting data

Introduction

Individuals vary; every person, every Fast Plant is different. These differences—differences that make life on earth so diverse—are measurable, and they allow us to uniquely describe each individual within a population.

But how do we describe a population? How do we characterize the population precisely and accurately to illustrate the variation?

We need to carefully collect information about the individuals in the population. That information, referred to as *data*, then needs to be organized and interpreted. Whether making observations or designing an experiment, it is important to consider

- what type of data to collect,
- how much data to collect, and
- how to organize the data.

When these steps are completed, you can effectively analyze the data and characterize the results.

Variation among plants can be as diverse as variation among classmates!

What Type of Data to Collect

The first step in collecting data is to decide **which data to collect** in order to characterize the population. Consider the information you need to measure, or do not need to measure, to generate a profile of the data.

Second, decide **how to measure the data**. Measuring the individuals precisely and accurately allows you to compare and summarize the data effectively and ultimately to draw conclusions about the population as a whole.

How Much Data to Collect

When designing an experiment it is important to determine the size of the population and the sample you are measuring. Consider whether it is necessary (or even possible!) to measure all the individuals in the population, or whether it would be more appropriate to measure a subset (or *representative sample*) of the individuals. Keep in mind that the number of individuals that are measured may determine how accurately the population is characterized. For example, with a population of 100 plants, how might your results differ if you sample 10 plants versus sampling all 100 plants?

How to Organize the Data

Two common ways of organizing data are in a graphical representation (frequency histogram), and a numerical representation (statistical summaries).

Graphical Representation of Data: Frequency Histogram

When several measurements are collected to create a data set, considerable variation may exist. By simply displaying these data as a set of numbers, shown on the right, relatively little information can be gained from them. A more effective way to organize data is in a *frequency histogram*.

48 Plant Height Measurements (mm)

To create a frequency histogram, data should be organized into groupings, called *class intervals*. For example, the data set of 48 plant height measurements (above) may be divided into classes that represent a 10-mm interval. Data are then grouped in intervals from 0–10 mm, 11–20 mm, 21–30 mm, etc., until the maximum value is reached.

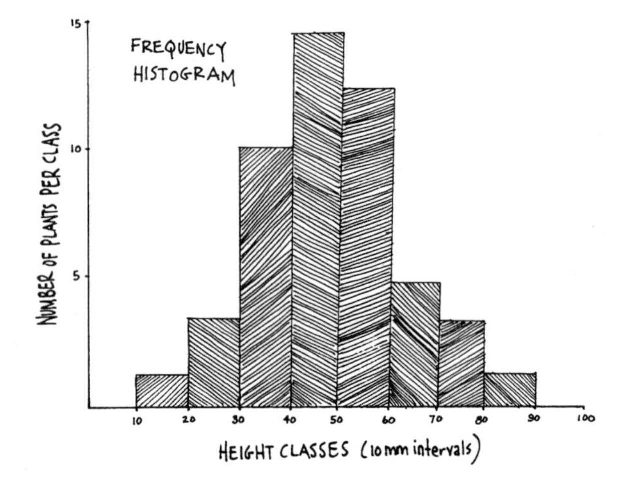

Class intervals can then be represented on the x-axis of a graph and plotted against the number of plants in each class on the y-axis.

Choosing the appropriate class interval size for the data is important; consider what happens to the profile of the data if the interval size increases or decreases. For example, how do the data look if the class sizes are reduced to 2-mm intervals or increased to 25-mm intervals?

Numerical Representation of Data: Statistical Summaries

Another way to efficiently organize the data is to use statistical summaries. Four useful summaries include:

n (number) = the number of data points.

$\bar{x}$ (mean) = the average of the data values; this is the sum of all values in the data set, divided by the total number of data points (n).

r (range) = the range of the data; this is the difference between the maximum and minimum values in the set.

s (standard deviation) = the average amount of variation from the mean ($\bar{x}$). Standard deviation can be determined on a calculator with statistical functions.*

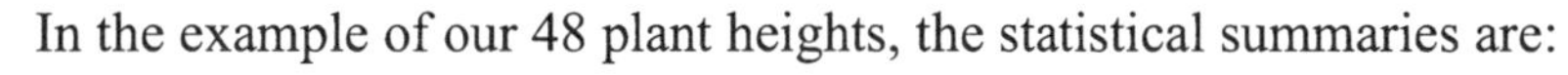

In the example of our 48 plant heights, the statistical summaries are:

number	n = 48 plants
mean	$\bar{x}$ = 47.13 mm
range	r = 66 mm
standard deviation	s = 14.27 mm

*Standard deviation is calculated as the square root of the sum of the squares of differences from the mean divided by n minus one (n–1). Standard deviation can be derived by:

1. Calculating the differences between each value and the mean (x).
2. Squaring each difference from Step 1.
3. Adding together all values from Step 2.
4. Dividing the sum from Step 3 by (n–1).
5. Taking the square root of the entire value from Step 4.

Interpreting the Data

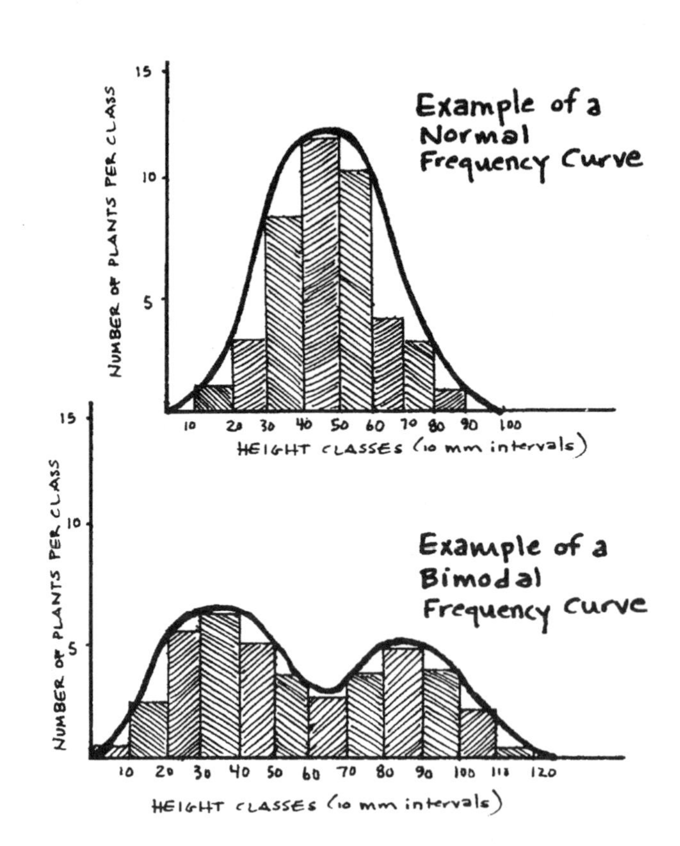

Once the data have been organized into frequency histograms and/or statistical summaries, the data can be interpreted. Frequency histograms portray the data graphically. A *normal distribution curve* or *normal curve* is commonly observed for many phenomena and is the basis for certain kinds of statistical summarization and interpretation. What does this graphical representation tell you about your population? Does your population exhibit a normal curve? Or do your data show multiple peaks, illustrating a bimodal or other distribution of data points?

Statistical summaries of the data portray the data numerically. What does the numerical summary tell you about the population?

For example, is the **number** (n = 48) a sufficient data set to draw conclusions about this population's plant height? Do more plants need to be measured? Could fewer plants be measured to characterize this population?*

*Detailed, specific answers to these questions come from an understanding of probability theories and statistics. Probabilistic statistics addressing the question of sample size and the significance of data as representative of a characteristic in a population are important aspects of scientific methodology that are not discussed in this Appendix.

Assuming **n** is sufficient, what do the mean, range, and standard deviation tell you about the variation in your population? The **mean** is the average value of the data set. When the values are distributed normally, the mean can be a useful number for summarizing or representing the set. However, what if the data set produces a bimodal frequency histogram or other distribution. Is the mean representative of the entire population?

The **range** depicts the difference between the maximum and minimum values. But what does it tell you about the rest of the data set? Is the range sufficient for understanding the variation of all of your data? Consider two sets of data. In the first set, all 29 data points are clustered around a mean of 24 mm with a minimum of 5 and a maximum value of 43, giving a range of 38. The second set is identical, except for 1 data point at 113 mm, giving a range of 108. How different are these two data sets?

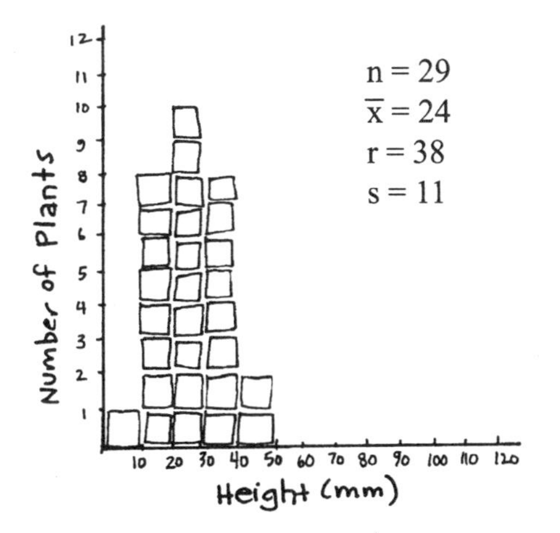

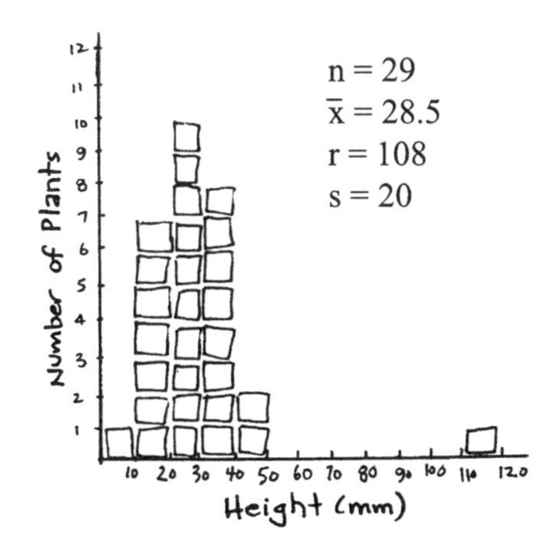

Standard deviation provides an indication of how the values of the data set are distributed around the mean. As the variation in a population increases, the value of the standard deviation increases and results in a wider frequency histogram. On a normal curve, one standard deviation is defined as the range that encompasses 66% of the data points that occur on both sides of the mean. If the standard deviation is low, less variation exists around the mean, producing a narrow frequency histogram.

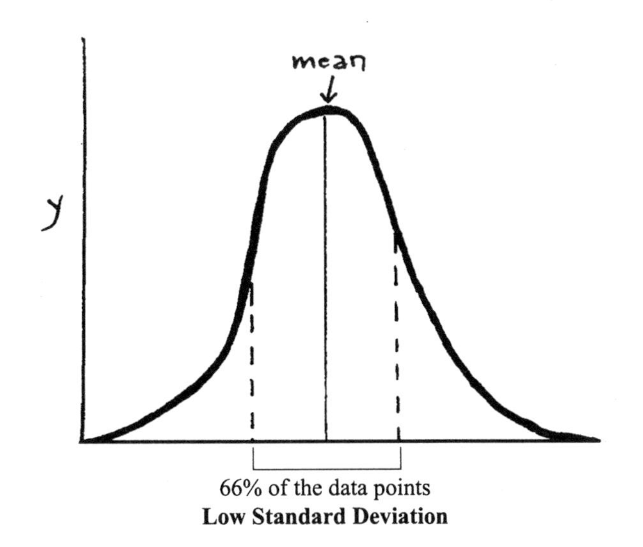

66% of the data points
Low Standard Deviation

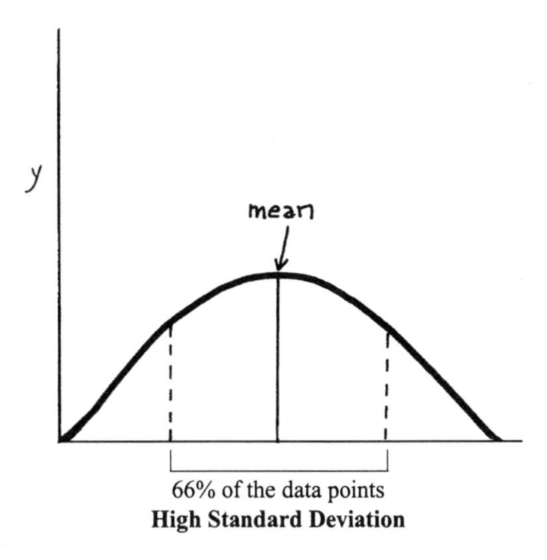

66% of the data points
High Standard Deviation

Conclusion

Characterizing a population requires careful planning and thoughtful interpretation. A well thought-out experimental design, including determination of what type of data to collect, how much data to collect, and how to summarize the data, is important to consider when setting up an experiment. With this planning you can provide useful information about the profile of a population that accurately reflects the variation between individuals. Through analysis and interpretation you will have a better understanding and be able to effectively communicate your results.

Additional Reading

D. Curran-Evert, "The Process of Scientific Discovery: How Certain Can We Be?" *The American Biology Teacher* **62**(4): 266–280, 2000.

E. A. Halpern, "Toward Scientific Literacy for Nonscience Majors: How to Approximate a t-Test by Graphical Means," *The American Biology Teacher* **62**(4): 276–281, 2000.

R. A. Johnson & K. Tsui, *Statistical Reasoning and Methods,* John Wiley & Sons, New York, 1998.

C. R. Rao, *Statistics and Truth: Putting Chance to Work*, International Co-operative Publishing House, Fairland, Maryland, 1989.

APPENDIX 3

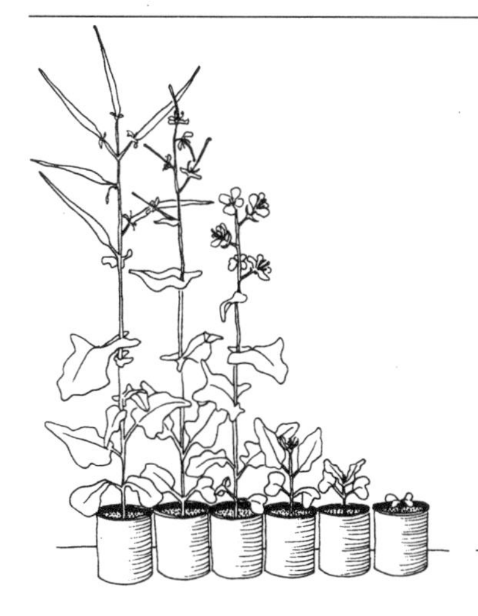

FILM CAN HAND LENS

Introduction

With a simple hand lens the world of microexploration comes alive. Pollen grains, developing embryos, petal abscission zones, bee setae (feather-like hairs), germinating seeds, flower pistils, and leaf hairs suddenly loom large. Access to another world presents itself.

A simple hand lens can be made with only an inexpensive lens, a film can, and a soda bottle cap.

Materials

- Translucent or clear film cans with lid (available at film processing stores)
- Plastic soda bottle cap
- Double convex plastic lens, 26 mm in diameter with 5X magnification and a focal length of 59 mm*
- Electric drill or drill press (recommended)
- 3/4-inch diameter wood bit with spurs
- 1/8-inch diameter drill bit
- Wooden dowel, 2.5 cm in diameter and approximately 5 cm in length
- Yarn or string (optional)

Preparation

1. Drill a 3/4-inch hole in the top of a soda bottle cap.
2. Drill a 3/4-inch hole in the bottom of a film can.
3. Optional: drill two 1/8-inch holes in the side of the film can, approximately 1/4 inch from the top of film can and 1/4 inch apart for inserting string for a necklace.

Construction

1. Drop the lens into the film can.
2. Set the bottle cap with the hole facing down into the open end of the film can.
3. Set the film can on a table; place a wood dowel into the bottle cap push the bottle cap down **firmly** until it sandwiches the lens into the bottom of the film can.

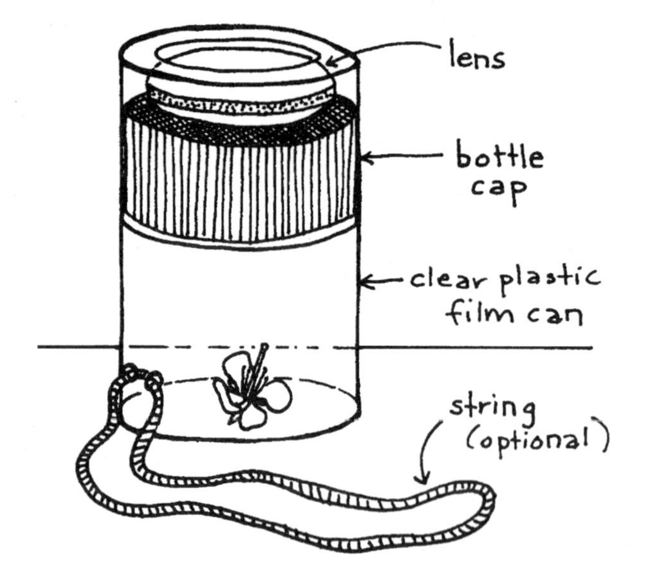

Tips and Suggestions

- A local film processing outlet or camera store is a good source for film cans. Ask them to save the cans for you, since cans are usually discarded.
- You can thread a string through two small holes made in the film can and wear the film can hand lens as a necklace.
- The hand lens can be used as a bug bottle by placing a bug in the can and closing the lid.
- You can melt or drill (13/16-inch drill bit) two larger holes in the side of the can to be used as access ports.

*This inexpensive lens can be ordered from Carolina Biological Supply Company (1-800-334-5551).

SCALE STRIPS, DRAWING TO SCALE, AND CALCULATING MAGNIFICATION

When observing something very closely, you can better understand the notions of size, scale, and magnification. **Scale Strips** help in handling small dissections and manipulating specimens for viewing under a microscope. They also provide a scaled reference to assist in drawing to scale and in calculating the magnification of a drawing.

Making and Using Scale Strips

Scale strips can be made by copying the black line master (page 137) onto a transparency sheet. The copied transparency sheet can be stuck, printed side down, to a laminating sheet or piece of clear contact paper, and then the individual strips can be cut out. Using the laminating sheet or contact paper as a sealer protects the printing from being pulled or rubbed off during use, but it is not a necessary step. Cut the strips out.

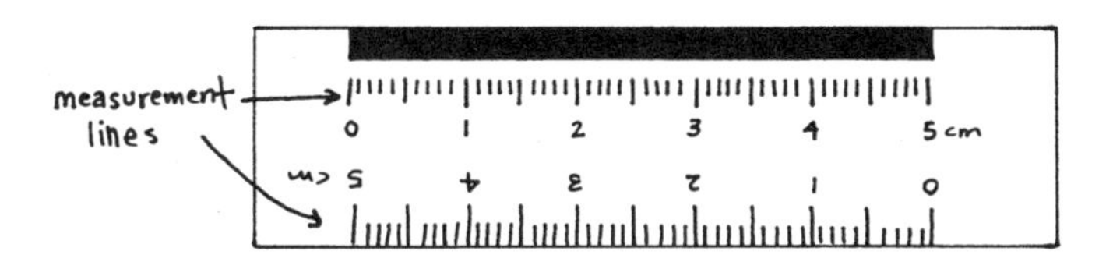

Once the strips are finished, they are ready to use. Begin by cutting a piece of 2-cm-wide clear adhesive tape to be about 3 cm long. Fold over about 0.75 cm of this piece of tape to make a tab. Stick this tabbed piece of tape to the scale strip, with the tab at the end of the strip.

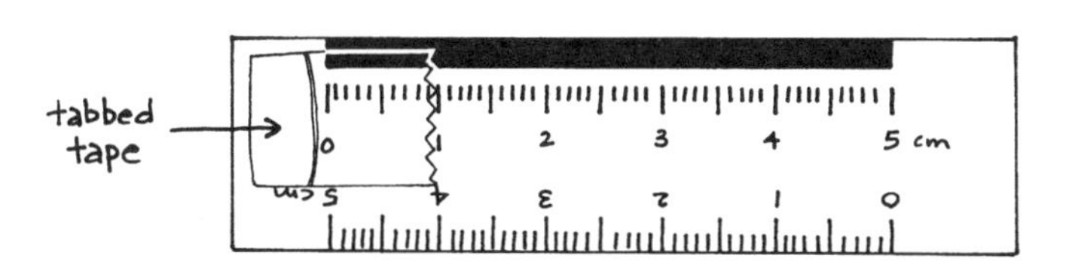

Cut a piece of clear double-stick tape. Place this piece near the top edge of the scale strip so that the end of the piece overlaps the tabbed piece of tape by a few millimeters.

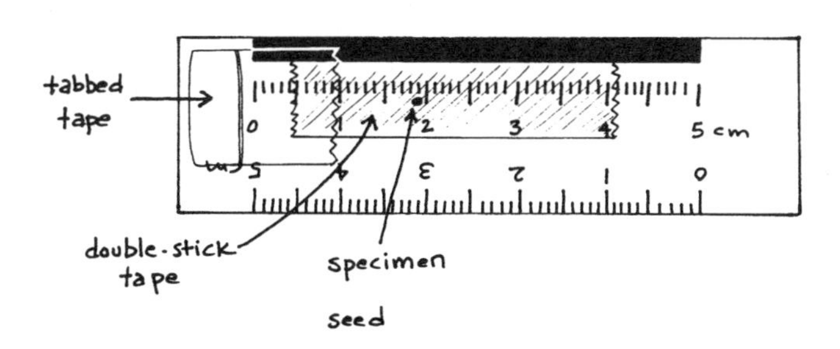

Specimens for dissection are placed on the double-stick tape. Once your specimen is in place, the specimen and strip can be placed under a dissecting microscope or a film can magnifier.

Once a dissection has been completed, the dissected specimen may be taped in a student notebook or removed from the strip by pulling up on the tabbed piece of tape. As this piece of tape is removed it will pull off the used double-stick tape and the strip will be ready for a new dissection.

Drawing to Scale

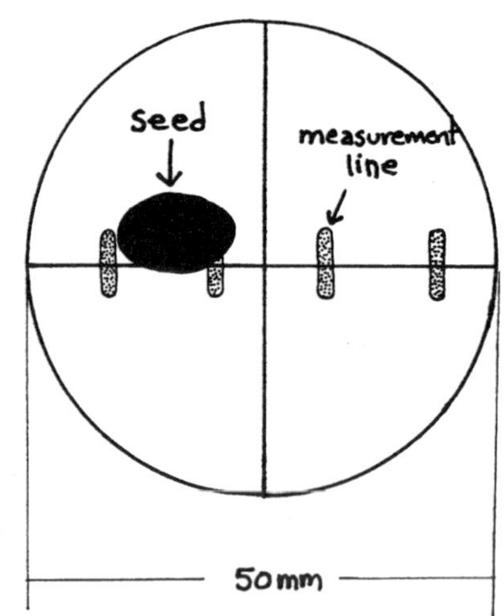

1. Place the specimen on a scale strip in such a way that the *measurement (ruler) lines* of the scale strip are observed together with the object.

2. On the sheet of paper on which you will draw a magnified picture of the object, first draw a portion of two adjacent magnified measurement lines as you see them through the magnifier or microscope. Be sure that the distance between the two lines is accurately represented.

3. On your sketch, draw a line (**scale bar**) between the two adjacent scale lines.

4. Be sure to note and record on your drawing the length the scale bar represents (e.g., 1 mm or 1 cm).

5. Next, draw your observed specimen as carefully as you can, being sure that your drawing is in the same magnified portion as the drawing of the scale bar. If you have kept the magnified drawing of the scale bar and the object in accurate proportion to each other you have drawn your object to scale.

Calculating the Magnification of Your Drawing

If you have drawn the specimen to scale, the calculation of the magnification of your drawing is straightforward.

1. With a ruler, measure and record the actual length of your drawn scale bar on paper.

2. The magnification of your drawing will be equal to the ratio of the length of the drawn scale bar on your paper in relation to the distance that the scale bar represents.

$$\text{magnification} = \frac{\text{actual length of drawn scale bar on paper}}{\text{distance the scale bar represents}}$$

For example, if the length of your scale bar represented 1 mm and the drawn scale bar on your paper actually measured 22 mm, then the magnification of your drawing would be calculated as follows:

$$\frac{22\text{ mm}}{1\text{ mm}} = 22\text{X magnification of drawing}$$

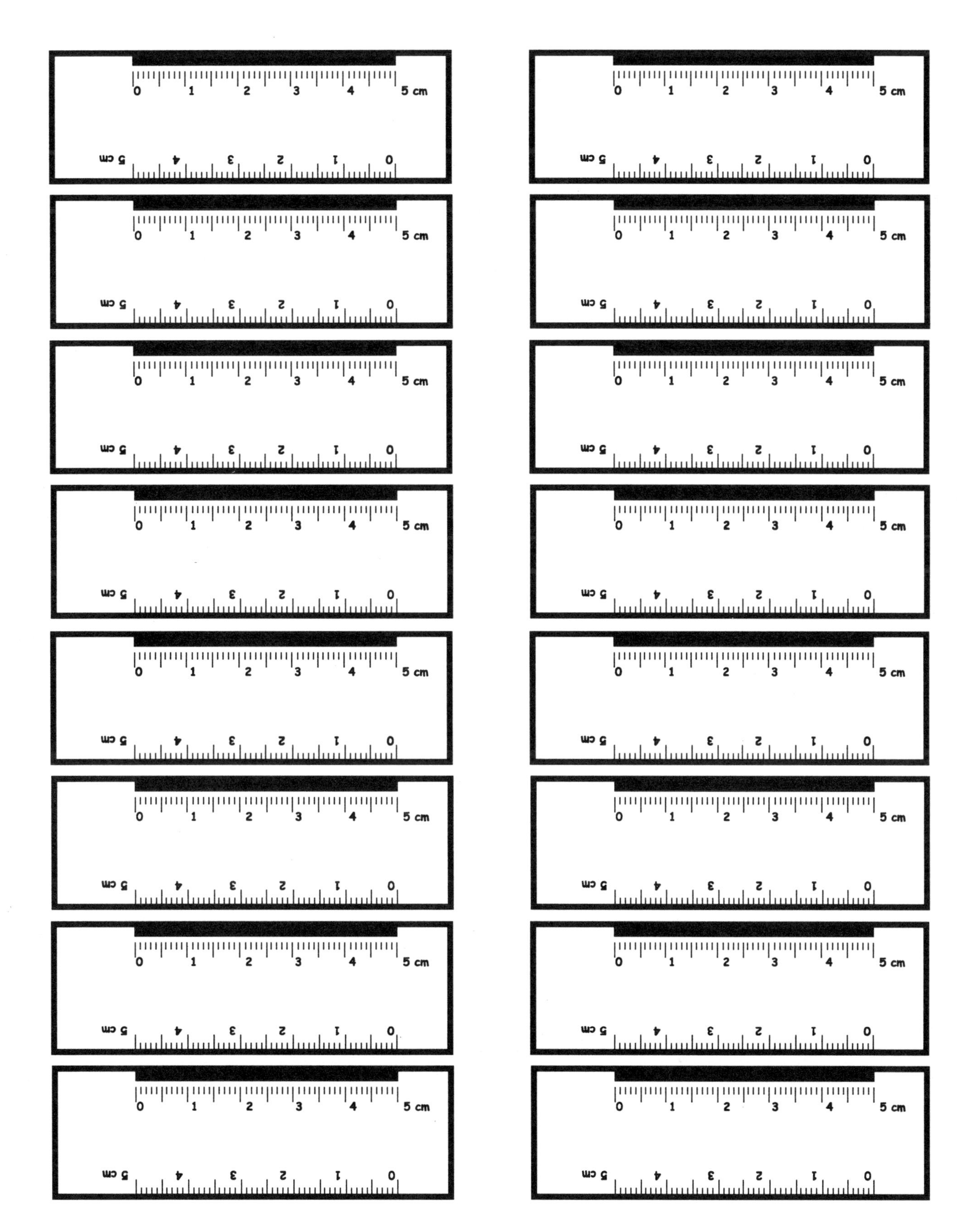

APPENDIX 5

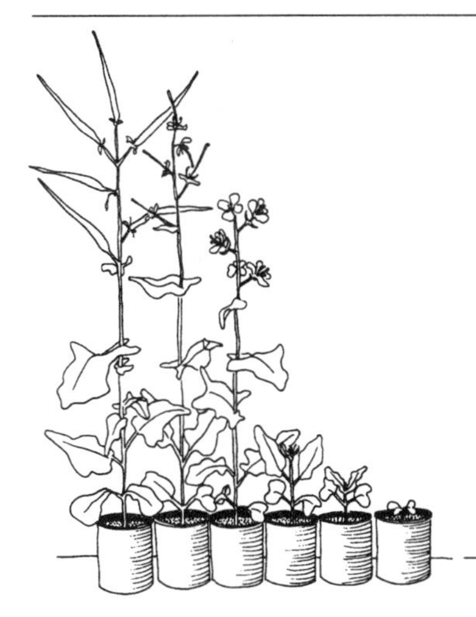

CLASSROOM VIGNETTES

INTRODUCTION

The following Appendix is a tale of three teachers. Each of the teachers wants to expose his or her students to science as an inquiry process of posing a question, designing an experiment, collecting and analyzing data, drawing conclusions, and presenting findings. They each want students to begin to understand the scientific inquiry process by grappling with experimental design, developing and modifying hypotheses, and bumping into and resolving ambiguities.

Each of the teachers approaches his or her task differently.

- Mr. L.'s high school biology students, having spent their first semester focusing on content, use their knowledge to enter the world of scientific research. Through the entire second semester, small groups will design, run, and analyze experiments and present findings from their research projects.

- For Ms. H., the scientific process is used as a way to introduce science and biology on the first day of her introductory high school biology class. Her message is that science IS experimenting. Quickly individuals are challenged to develop their own explorations.

- Mr. B. uses Fast Plants to walk his 7th grade students through the scientific process, challenging them with the inevitable ambiguities that emerge. Classroom discussion in concert with individual replicates of a class experiment form the basis for this three-day experience.

In each classroom, Fast Plants are used to inspire student-led explorations. This model organism lends itself to open-ended inquiry in the classroom because of its simple growth requirements, low cost, petite size, and rapid life cycle.

Mr. L. Ninth Grade

The first science class after Christmas is unusually full of energy. Mr. L.'s students are informally surveying their classmates to see which research project they are interested in doing. Mr. L.'s second semester classes are not unlike other introductory biology classes where the topics of genetics, protein synthesis, evolution, taxonomy, and the like assume major curricular importance.

However, Mr. L. wants to go further.

He wants to design an educational setting, in the confines of a freshman introductory biology curriculum, where his students can demonstrate their ability to apply the information and experiences they gained in the first semester to a new educational situation in the second semester. The students will build on their first semester experience of growing Wisconsin Fast Plants to model the actions and thought processes of a scientific researcher.

The chatter in the room today tells Mr. L. that these additional challenges of designing and conducting research utilizing the Fast Plants have more than a few students a little nervous. He collects the vacation assignments—write a question and design a controlled experiment to attempt to answer your question—while avoiding the pitfall of discussing them. Written feedback will be provided on their ideas in the very next class period. The feedback is the first in a number of "scientific dialogues" that will take place between student and teacher throughout the semester where students can reference and build off of their Fast Plant experiences as well as any other relevant materials that have been covered in the scope of the biology course. Class then focuses on the new topic at hand, genetics.

The next day, students receive their "corrected" assignments and quickly notice that there are many questions and comments written on their papers: How are you going to control this variable? What is your sample size? How will this be monitored on the weekends? Do we have the equipment that you would like to use? Mr. L. quells the chaos by informing the students that he will be glad to meet individually to spend time discussing their specific concerns. Many students set up appointments to see him during their free time—study hall, lunch, before school, and after school.

Even with individual consultations, many students remain unsettled. These students, like many others, are used to traditional laboratory work where the teacher hands out a lab sheet and the student follows the directions to the prescribed outcome. However, Mr. L. has subtly been preparing his students for this experience by discussing the experimental design of these "cookbook" labs and encouraging his students to come up with a laboratory procedure for a lab before he hands out the laboratory sheet. Up to this point, many just considered this an additional component of their laboratory work. Now they are putting those critical thinking skills to work on their own, very real experimental design problems.

The next two weeks are very hectic. Mr. L. has entered into five- to ten-minute scientific dialogues with many of the student groups, frequently sending the group away with additional questions, ideas to consider, and background information to research. He is very sensitive to the students' frustration level and is recording the

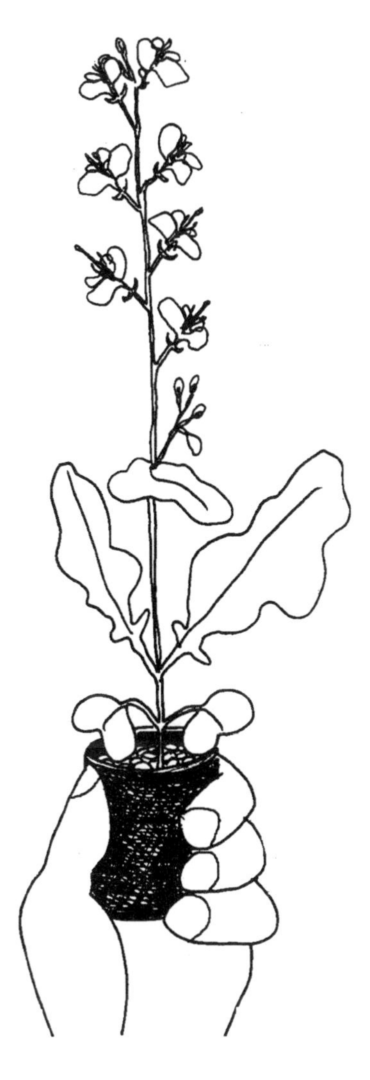

progress or lack of progress for each of the groups. These dialogues are a key component of the evaluation process. By the end of the second week, a fair number of the groups are given permission to conduct the research and Mr. L.'s role shifts from teacher to research director. As research director, Mr. L. monitors the equipment use, room use, and safety issues. He also requests periodic verbal progress reports. These reports provide opportunities to engage students in conversations regarding their projects, revealing their understanding of their work.

Mr. L. encourages collaboration with students in different sections who are conducting similar experiments. Often students search the Internet for other schools conducting similar studies. With these collaborators, students can discuss particular challenges and surprising results, and they can sometimes pool data to address the issue of sample size and number of trials in their studies. This process in and of itself is educationally rich. Mr. L. has observed students discussing issues of controlling variables, defining operational terms, and collecting and analyzing data with fellow students. A few years ago, Mr. L. arranged for his students to link up with two other schools conducting the same type of project, one in Connecticut and the other in California.

The process of refining the question and experimental design takes some student groups longer than others, which serves to reduce the immediate demand on materials and laboratory space. So, with many of the projects underway, Mr. L. devotes his energies to the students who haven't started their research. Occasionally class time will be devoted to a research team that is having particular difficulty with their research and the class brainstorms solutions and ideas. By the end of week 4 every research group is making progress. The projects range from using time-lapse photography to record the life cycle of *Brassica rapa* (Fast Plants), to determining the effects of soil composition or plant hormones on growth and development and developing age-appropriate curricular material for use in the preschool and elementary classroom. In the role of research director, Mr. L. keeps tabs on all of the groups.

Toward the end of the third quarter the students are provided with some class time to set up and enter data on computer spread sheets. Many utilize this time to begin writing the initial sections of their laboratory report—introduction, background information, materials, procedure, safety—in preparation for their presentations during the final exam week in the fourth quarter. Mr. L. fields many of their questions by asking students how they handled a particular issue on a past lab or by directly relating their question to their experiences writing the Fast Plant laboratory report in the first semester.

Time management issues always seem to lie just below the surface throughout this project. The students are aware that these plants take about 38 days to complete their life cycle and that the laboratory reports are due toward the end of the fourth quarter, some nine weeks away. But the unexpected does arise. The beginning of the fourth quarter, a full seven weeks into the project for many of the students, reveals the rare research group facing a perceived insurmountable obstacle—dead plants or lost data. Always cognizant of the students' frustration level, Mr. L. provides these groups with options to conduct "mini" research projects on germination and cotyledon development, which can be completed in roughly ten days. While these students engage in intensive discussions with Mr. L. to define their new questions, they are still accountable for demonstrating ownership of their new research focus. They must clearly articulate the methodology utilized, background information, and experimental design. These groups are now faced with writing two reports, one reflecting on the progress and end point of their initial research, and the second focusing on their "mini" research project.

With all of the groups on track and just two weeks to go before the final exam, Mr. L. distributes a Fast Plants presentation scoring rubric and sets the due date for the final laboratory report. Most of the students find that the process of development and revision of the more traditional Fast Plant lab report in the first semester was sufficient to eliminate the mystery of writing this final report. The preparation of the oral presentation, on the other hand, is quite a different story. In an effort to reduce student anxiety, Mr. L. shows videotapes from previous years and spends a period or two of class time discussing presentation details such as eye contact, reading from note cards, speed of the presentation, size of the visuals used, and body language. This seems to relieve at least some of the jitters.

The presentation day arrives. The school's professional library is the location for this symposium of Fast Plants research. The audience for these presentations includes parents, administrators, and fellow student researchers. Some of the students have spent their free periods over the past couple of days setting up computers for PowerPoint® slideshows, while others have arranged for slide projectors, easels, or overheads to aid in their presentations. Mr. L. arrives with refreshments and a printed presentation schedule. Mr. L. is especially interested in the students' use of vocabulary and how they field audience questions based on their research.

It has become very clear by the students' performances that they have gone beyond just completing another course requirement. They have integrated and internalized many of the major themes in science: asking questions, controlling variables, understanding the limits of your experiment, communicating ideas, and showing biological connections. In essence they have satisfied Mr. L.'s goals of having students doing science instead of having science done to them.

Ms. H.
Ninth Grade

The first day of Biology 1 at Medfield High School is not what the students expect. Ms. H. wants to establish three things: class is fun, curiosity is good, and science is experimenting. Instead of the typical introductions including expectations and passing out of textbooks, Ms. H. introduces herself and announces that since they are about to embark on a year of biology, they need to start with the question, "What is biology?" Students suggest definitions such as "the study of life," "looking at living things," and "studying animals." Ms. H. records their responses on the board.

As life seems to be a central theme in their answers, Ms. H. then asks, "What is life?" Again a list is generated on the board as the students begin to define a living thing: it breathes, it moves, it is made of cells, it makes noise, and it needs water.

"If biology deals with living things," suggests Ms. H., "we need to get a better idea of what 'living' really means. As a matter of fact, there is a living organism right next to each of you at your lab benches."

"Huh?" A few startled students begin to glance about some with some trepidation.

"No kidding," Ms. H. continues, "there is a living organism within one meter of each of you!" This assertion causes great excitement as the students assume that some sort of small, creepy thing must be approaching them. The students begin to search for the creature.

Prior to students entering the room, Ms. H. has taped a Fast Plant seed near each student station. Using a very small piece of tape, the seeds are attached to the underside of the lab bench, on the side of a table leg, or on the edge of a seat.

Most students are searching for something bigger, so Ms. H. must give some clues until someone eventually finds a seed. Once they know what they are searching for, the others find their seeds rapidly. As students protest that a seed shouldn't count as a living thing, Ms. H. returns their question, "How could you determine if a seed is indeed living?"

After some discussion, someone usually suggests that they could do an experiment to determine if the seed is alive. Ms. H. asks everyone to write down what type of questions would have to be answered in order for them to determine if the seed is really a living thing. After a few minutes, Ms. H. asks them to write down what they already know about seeds and plants.

Ms. H. challenges the students one step further, "Our question is 'Is a seed a living thing?' Given what you know and what you need to know to answer that question, consider how you might set up an experiment to begin to address this issue." She shows them a tray of available supplies that includes film canisters, microcentrifuge tubes, paper towels, Fast Plant seeds, and water.

"Develop an experimental design and write down the procedure." Except for encouraging them to keep the experiment relatively simple, Ms. H. does not interfere with their designs—these are *their* experiments.

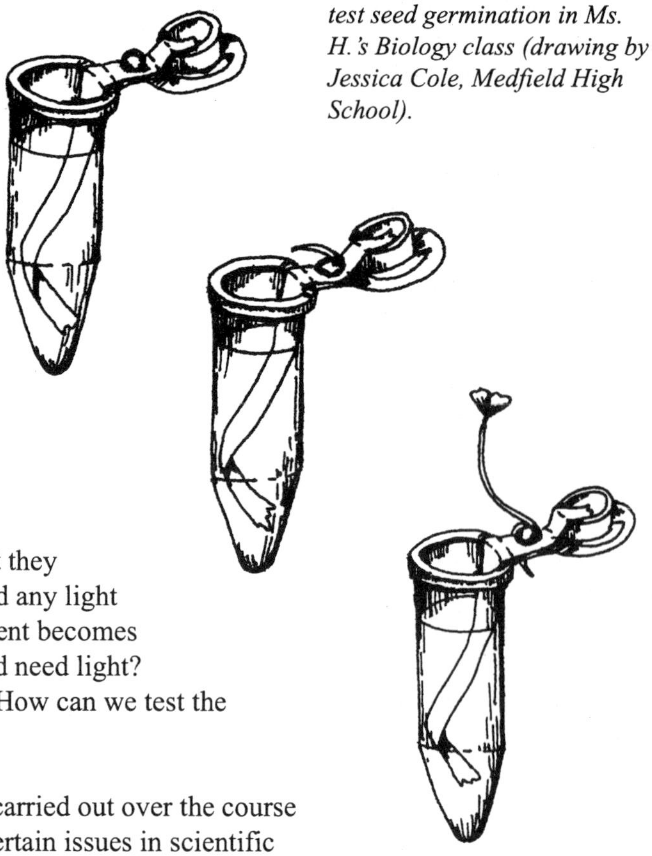

A typical experimental setup to test seed germination in Ms. H.'s Biology class (drawing by Jessica Cole, Medfield High School).

The students set up their experiments and write down their predictions of what will happen. In general, students figure that if they want to prove that the seed is alive, they must get it to grow. There are all sorts of variations on the idea that if they provide water, the seed will grow. The experiments are left overnight. The next day, most seeds are undergoing some change such as swelling or a tiny root emerging. The students continue to take observations for a couple of days.

The class discusses the results of the experiments, what they learned, any conclusions they have, and if they can shed any light on their main question, "Is a seed alive?" This experiment becomes the jumping-off point for several questions: Does a seed need light? What temperature is best for growth and germination? How can we test the toxicity of different compounds? Do seeds need soil?

Experiments based on the students' questions are then carried out over the course of the next few weeks. Some experiments emphasize certain issues in scientific research. For example, the issue of using controls in experimentation—when controls are necessary, what constitutes a good control, and what a control tells us—are all issues that arise naturally as the class designs and carries out temperature and light experiments. Bioassay experiments provide a good venue for developing graphs and tables that simply and precisely convey an experimental result. Other experiments provide opportunities to explore topics such as the scientific method, variables, interpreting results, and handling data.

On that first day of biology class at Medfield High, students are not introduced to their class but instead they are introduced to science. They are not told the expectations of behavior and assignments, but rather are shown the expectations of their involvement. They see that the emphasis for their learning is not books and information but experimentation and thinking. And that is precisely the message that Ms. H. intends to send.

Mr. B.
Seventh Grade

"OK! Let's rapa it up!"

"Arrrgh! Mr. B.! That is soooo weak!!!"

"You have ten minutes! Get with it! Make sure you have labeled the fronts of your film cans with the date, hour, and initials."

Mr. B.'s seventh grade science class will be using Fast Plants, a variety of *Brassica rapa*, to learn about plant growth and development.

But first he will use the three-day-old seedlings to introduce the kids to the world of science inquiry. He uses the inquiry process to help students sharpen the critical thinking skills that he emphasizes throughout the course. This will be the first of several student research experiences in the year.

A group of three students approaches Mr. B. with their film cans in hand. "Mr. B., we have the seedlings stuck to the sides of the film can by their hypocotyls…" Scott glances at the drawing on the board, "—their cotyledons. Do we need to cover the open end with a lid?"

Avoiding giving an answer, Mr. B. coaxes the students to think further. "Well, what do you think?"

"If we don't cover it, it could dry out?"

"That's a thought.…Katie, what do you think?"

"Wouldn't we also want a cover on the film can to eliminate light?"

"Oh yeah!" The light bulb goes off in Scott's head. "We want to remove light as a variable!"

Jackie joins in, "But I thought plants needed light to grow!? Won't they die without light?"

"That's a good question!" Mr. B. adds with a grin.

It seems simple enough. The students are setting up one of Mr. B.'s favorite activities: The Crucifer Cross. In this experiment, students cut four three-day-old Fast Plant seedlings at the soil line leaving the stem (hypocotyl) and seedling leaves (cotyledons) intact. A gravitropism chamber is then constructed by placing four moist paper towel strip wicks along the inner side of a film can so that there is one wick each along the length of the top, bottom, and both sides of the can when it is placed on its side. The cotyledons of each seedling are then stuck to each wick so that the water on the wick holds the seedling in place.

He turns to the class and addresses them as a whole. "Remember! When you are done, you need to make a prediction. Copy the frontal view into your lab notebooks. Then draw what you expect the hypocotyls to look like by tomorrow. Stick your necks out! Make a prediction!"

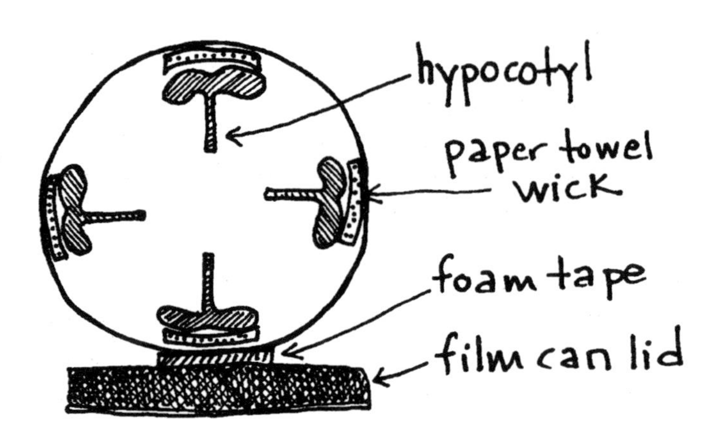

Joann raises her hand. "What do you mean? Like, what will the stems do?"

"Yes. What do you think the four hypocotyls will look like when you open this tomorrow?"

Bobby interjects, "You mean like all limp and dead?" His friends snicker.

"If that's what you think, then write that down. But it won't be an easy way out because you must back up your prediction with some solid reasoning."

"No! I'm not slackin'. Seriously! We killed them! They are dead, so—"

Mr. B. cuts in, "Is that an observation, or an inference?"

A sprinkling of kids chimes in, "Inference!!!"

"Why?" He scans the horizon for a hand in the air. "Jaime?"

"Because he's saying they are dead, but doesn't KNOW they are dead. He doesn't have any data...observations to support that."

"But," Bobby protests, "we cut them down. Slashed them above their roots! They are dead!"

"OK! OK! Hold on to those thoughts. Journal on it if it helps you sleep tonight. So, maybe we might want to clarify. 'Assuming they are still alive, I'd expect...' "

Jason blurts out, "OH! That maybe they'd bend toward the back?"

Mr. B. continues, "OK, let's run with that. How else could these 'stems' respond? Don't tell me why you say it, just give me some ideas."

Sarah's hand shoots up, "They could bend downward?"

"OK." Mr. B. draws dotted lines down, representing hypocotyls that are bent down.

"What else? Billy?"

"Well, they could all stay straight, or maybe bend up?"

"OK. So we have some options here: bend up, bend down, stay straight, toward front or back."

Mr. B. announces one last time, "Make sure you draw your predictions for tomorrow. Then write a brief passage telling why you think they will do this! You cannot open your canister unless you've made a prediction. No peeking and no talking to other classes!"

Day 2

Seventh hour seems to come more quickly today. Students stream in a few at a time. The students are pumped to open their canisters. Some fidget in anticipation. Others feign disinterest.

"OK, everyone has made a prediction! When you open your film can, you should draw the Fast Plants the way you see them. Go do it!"

Things get quite loud on days like today. Geoff, seated near the front, has some interesting results. His Fast Plants have managed to stay adhered to the side. That is crucial! Mr. B. asks Geoff to make a chart on the board, as his peers provide categories for grouping their data.

"Start with yours, Geoff." Geoff approaches the front and draws a picture of what is inside his canister. The two horizontal hypocotyls have curved upwards. The one pointing down has arched 180 degrees, bending it into a U-shape. Only the hypocotyl originally pointing upward has remained unchanged.

Geoff asks each of his peers what they got for results. As the numbers are tallied up, it is clear that a trend can be seen in the data with 19 students reporting "up," one reporting "down," two reporting "same," and six reporting inconclusiveresults. Now they need to draw some conclusions.

"So! What do you think? What should we do with these observations?"

Lynn starts things rolling, "Trina, how did you get 'down'?"

Trina looks inside her canister and replies, "Well, the one on the left is kinda, a little bit, tipped down." A few kids snicker.

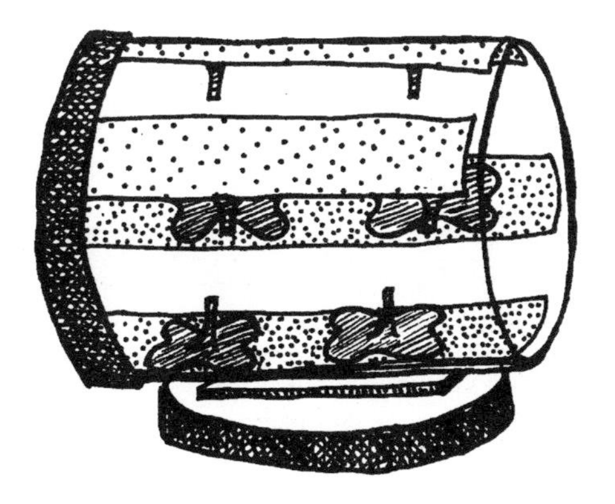

Mr. B. interrupts, "Hey! Watch it! It's not her fault if we didn't develop some better criteria! What do we mean if a canister is recorded as 'up' or 'down'?"

An outburst ensues, as kids clarify with neighbors just what they had meant by "up" or "down." As things quiet down, a self-appointed spokesperson blurts out, "To call it an 'up,' all hypocotyls had to be pointing up or at least three of them!"

Mr. B. asks for clarification. "At least three? So if one was pointing down…"

John cuts to the chase. "No, if the three, besides the bottom one, were pointed up, then we called it 'up.' So, like, Katie's bottom one tipped over, but we can still count hers as 'up' because the other three bent up."

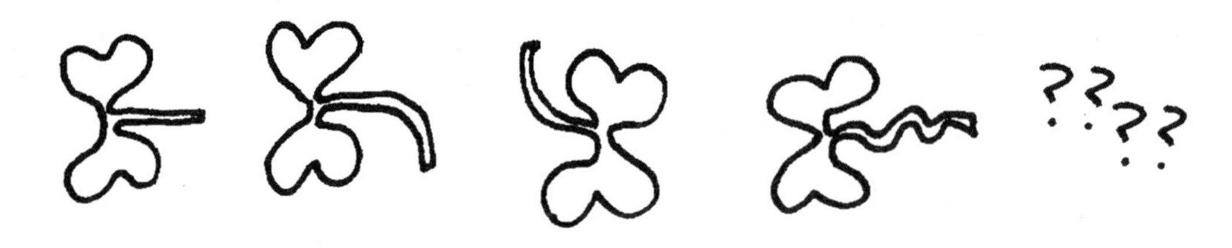

Mr. B. attempts to summarize: "So, you all agree that to score yours as 'up,' you should have all except the bottom one curved up?"

"YES!"

Steve still isn't satisfied. "If some fell off I think we should ignore them. We should count the can as 'up' if the ones that stayed stuck curved up."

Cindi nods in agreement. "Mine was 'inconclusive' because three had fallen off but the one that stayed stuck definitely curved upwards.

"So what are our criteria?" Mr. B. pushes.

Steve is on to it. "If any one curves upwards, it counts as an 'up.' And it has to *curve* so the one that started out pointing up doesn't count." Most people nod, satisfied.

"So with those criteria, there are 21 'up.' We can at least say that there's a consistent *tendency* toward bending up. Right? OK…is that what you predicted?"

The noise level rises again.

"Given our class data—and I'll add that we've basically seen the same thing throughout the day—what do you think is going on?"

A hand flies up.

"Wait, I want you to write! Assume now that we are seeing a tendency toward curving up. How would you account for this? What's causing this? Write it in your journal for tomorrow."

Day 3

Mr. B. knows that a challenging discussion faces the class today. Developing hypotheses to account for unexpected results is a challenge in any research lab, but precisely the kind of stick-your-neck-out creative thinking that middle school students generally are not encouraged to do either academically or socially.

"OK! Ideas? What's going on here? Why would these stems bend up?" Even though he isn't sure why the plants did this, Mr. B. has some ideas, but he is after something else. He has seen this before; kids have a hunch, a feeling—call it a hypothesis if that makes you feel good. But they will reserve comments until someone breaks the ice. Looks like he'll need to push them today.

"Carlos?"

"Huh?"

"C'mon, give me some idea, a guess, about why these plants are bending up."

"You mean a hypothesis?"

"Sure, if you want to call it that. Give me a hypothesis."

Ashley adds, "Ummm…isn't that an educated guess?"

A number of students burst out laughing. Carlos laughs and goes on, "Yeah, that's my point! I don't really know much about plants, so I don't think I'll give the right answer."

Mr. B. smiles. "Is that what experimentation is about? Getting the right answer? Carlos has said something that I am certain many of you were thinking—'What's the right answer?' or 'I'm gonna say something stupid!' And that is the difference between traditional cookbook labs and real science. In cookbook labs everything has been done a thousand times and, if done correctly, there is a right answer. But this…this is real science. The plants bent correctly—what happened, happened. Now we need to think about that, and in order to really come up with a good idea, we are going to need to think creatively—take chances, engage your brains, be messy, make mistakes, and make lots and lots of observations. It's not about answers so much as it is about questions." He goes on, "So, again, why do these plants curve up? Jamie?"

"Maybe they are looking for water?"

"OK…" Mr. B. writes this on the board. "Water…any others? Carlos?"

"Maybe gases? Like oxygen or carbon dioxide?"

"OK…" The dance continues as the students generate a half-dozen possible variables that could be causing the plants to curl. The tension in the room decreases in the room as kids begin to realize that this is a safe environment to share.

After reviewing the potential causal factors with his students Mr. B. takes it a step further. "OK, now I want you to think about which of these factors you feel is most likely to cause the curving of the hypocotyls. Write this down in your lab notebooks. Then, I want you to write a brief statement telling

how you would test to see if this is indeed the factor that causes the plants to curl. You've got two minutes—no talking."

Mr. B. gives them a few minutes to jot down some thoughts. As the students begin reaching some closure, he announces: "Now! Go to your lab stations and talk amongst yourselves. What variable would your group like to experiment with and why? This should be something that we could do with the resources we have here—the film can setups. First-come, first-served, no repeats. You've got seven to ten minutes, and then we've got to share ideas."

After ten minutes, Mr. B. announces that the students will need to finalize things within their groups, then give an informal explanation of what they are thinking about doing. As students share their experimental designs, their peers, within and between groups, offer advice and criticism. By the time they hit the last group, there is just enough time to assign a journal entry: explain, in your own words, what your group would like to do for an experiment. Use the terms "control" and "experimental variable." Predict what will happen and explain what kind of data you will collect to help support your claims.

Overall, the discussions have been broad ranging, the experimental results are messy, and the conclusions ambiguous. But Mr. B feels that the students are gaining a more realistic understanding of and appreciation for the inquiry process of science. And, scientists or not, this will serve them well later in life.

APPENDIX 6

SCIENCE STANDARDS

National, state, and district science standards are encouraging teachers to incorporate new modes of inquiry-based learning as well as providing guidelines for specific subject areas.

The focus of this manual is inquiry-based science. For example, the activities and inquiry launch pads meet and exceed the following high school science inquiry performance standards*:

- When studying science content ask questions suggested by current social issues, scientific literature, and observations of phenomena, build hypotheses that might answer some of these questions, design possible investigations, and describe results that might emerge from such investigations.
- Identify issues from an area of science study, write questions that could be investigated, review previous research on these questions, and design and conduct responsible and safe investigations to help answer the questions.
- Evaluate the data collected during an investigation, critique the data-collection procedures and results, and suggest ways to make any needed improvements.
- During investigations, choose the best data-collection procedures and materials available, use them competently, and calculate the degree of precision of the resulting data.
- Use the explanations and models found in the earth and space, life and environmental, and physical sciences to develop likely explanations for the results of their investigations.
- Present the results of investigations to groups concerned with the issues, explaining the meaning and implications of the results, and answering questions in terms the audience can understand.
- Evaluate reports in the popular press, in scientific journals, on television, and on the Internet, using criteria related to accuracy, degree of error, sampling, treatment of data, and other standards of experimental design.

As noted in the Introduction, the levels of detail and complexity in this manual are balanced with the high school classroom as the midpoint audience. By design, the activities can be scaled down to meet many of the content and inquiry science standards for middle school, and scaled up to provide rich content and inquiry investigations for undergraduate classrooms. The following page shows points of alignment of Fast Plants activities in this manual to the National Science Education Content Standards.

*Source: Wisconsin Department of Public Instruction, Madison, Wisconsin.
**Source: United States National Academy of Sciences.

National Science Education Content Standards

✓ indicates that there are concepts and activities in this manual that address the given standard.

Unifying Concepts and Processes

K–12

✓ Systems, order, and organization
✓ Evidence, models, and explanation
✓ Change, constancy, and measurement
Evolution and equilibrium
✓ Form and function

Science as Inquiry

K–12

✓ Abilities necessary to do scientific inquiry
✓ Understandings about scientific inquiry

Life Science

K–4

✓ Characteristics of organisms
✓ Life cycles of organisms
✓ Organisms and environments

5–8

✓ Structure and function in living systems
✓ Reproduction and heredity
✓ Regulation and behavior
✓ Populations and ecosystems
✓ Diversity and adaptations of organisms

9–12

✓ The cell
✓ Molecular basis of heredity
✓ Biological evolution
✓ Interdependence of organisms
✓ Behavior of organisms
✓ Matter, energy, and organization in living systems

Science and Technology

✓ Abilities of technological design
✓ Understandings about science and technology

History and Nature of Science

K–12

✓ Science as a human endeavor

5–8

✓ Nature of science
✓ History of science

9–12

✓ Nature of scientific knowledge
✓ Historical perspectives

W. M. Becker, L. Kleinsmith & J. Hardin, *The World of the Cell* (4th edition), Benjamin/Cummings Publishing Company, New York, 2000. A current and approachable college-level cell biology text.

N. A. Campbell, J. B. Reece, & L. G. Mitchell, *Biology* (5th edition), Benjamin/Cummings Publishing Company, New York, 1999. An excellent introductory college-level biology text book.

K. Esau, *Anatomy of Seed Plants* (2nd edition), John Wiley & Sons, New York, 1977. The long-standing college text on plant anatomy.

R. A. Johnson & K. Tsui, *Statistical Reasoning and Methods,* John Wiley & Sons, New York, 1998.

P. H. Raven, R. F. Evert & S. E. Eichorn, *Biology of Plants* (6th edition), W. H. Freeman and Company, New York, 1999. A thorough coverage of college-level plant biology.

A. Sugden, *Longman Illustrated Dictionary of Botany*, Longman-York Press, Naperville, IL, 1984.

L. Taiz & E. Zeiger, *Plant Physiology* (2nd edition), Sinauer Associates, Inc., Sunderland, MA, 1998. A current college text covering the field of plant physiology.

P. Williams et. al., *Bottle Biology: An Idea Book for Exploring the World Through Plastic Bottles and Other Recyclable Materials* (2nd edition), Kendall/Hunt Publishing Co., Dubuque, IA, 2003.

P. Williams et. al., *Exploring with Wisconsin Fast Plants* (2nd edition), Kendall/Hunt Publishing Co., Dubuque, IA, 1997.

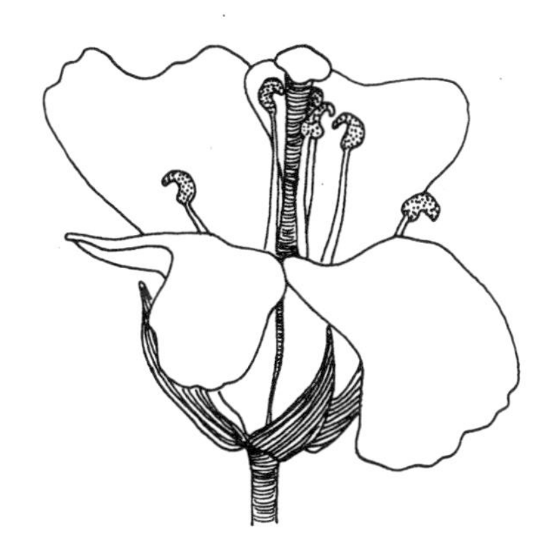

GLOSSARY

Abscission — the dropping off of leaves, flowers, fruits, or other plant parts.

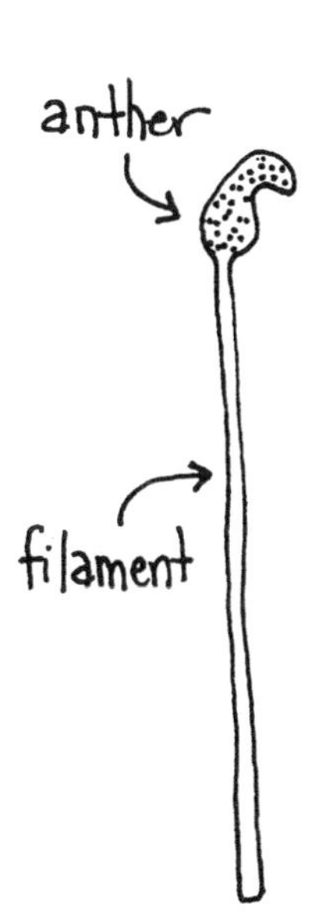

Abscission zone — the area at the base of the leaf, flower, fruit, or other plant part containing tissues that play a role in the separation of the part from the plant body.

Anther — the pollen-bearing portion of the stamen.

Anthesis — the shedding of pollen from the anther.

Apical meristem — the meristem at the tip of the root or the shoot.

Asexual reproduction — any reproductive process that does not involve the union of gametes.

Auxins — a class of plant hormones that are involved in many aspects of plant growth and development.

Basal cell — a cell that anchors the developing embryo to the ovule.

Carpel — the structure in fruit-producing plants that encloses one or more ovules.

Coevolution — the simultaneous evolution of adaptations in two or more populations that interact so closely that each is a strong selective force on the other(s).

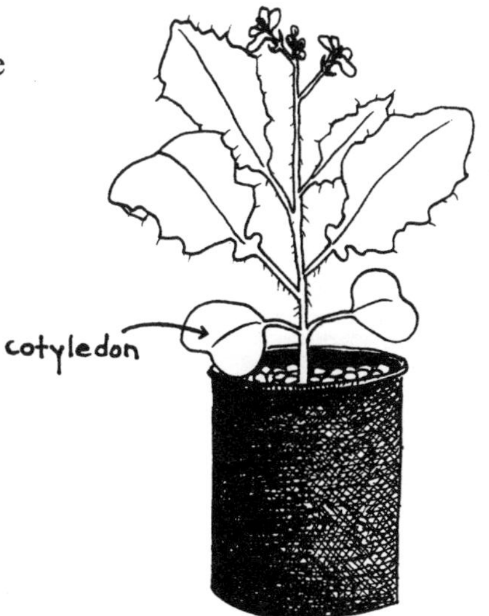

Coleoptile — protective sheath on a monocot seedling shoot.

Cotyledon — seed leaf, first to appear after germination in Fast Plants. It serves as a food source (energy) until true leaves form.

Cross-pollination — the transfer of pollen from the anther of one plant to the stigma of a flower on another plant.

Cross section — representation of a cut made perpendicular to the main axis of an object.

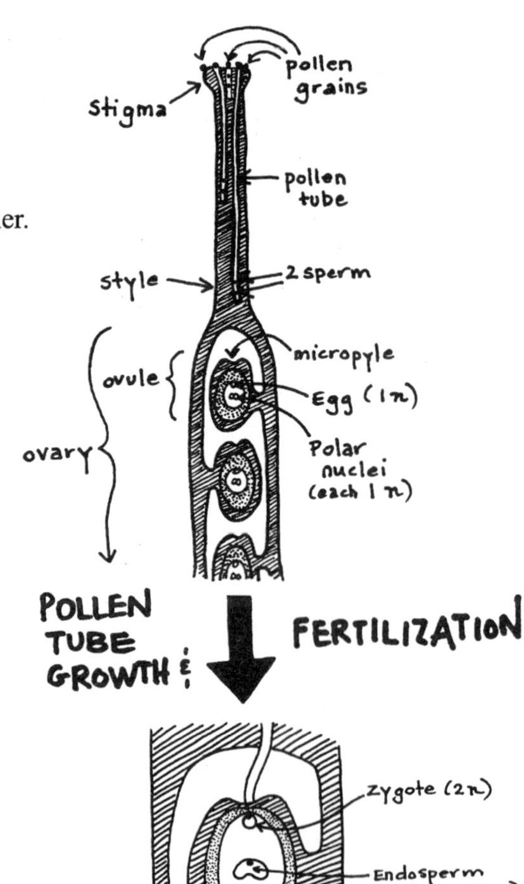

Dap — days after pollination.

Das — days after sowing.

Dehiscence — the release of pollen to the surface of an open anther.

Dicot (dicotyledon) — a plant whose embryo has two cotyledons.

Double fertilization — the fusion of the egg and sperm, and the simultaneous fusion of a second sperm with the polar nuclei resulting in the endosperm.

Embryo — a young plant before the start of germination.

Embryogenesis — the development of an embryo from a fertilized egg or zygote.

Endosperm — the tissue that forms in the ovule after fertilization as food for the developing embryo.

Epicotyl — embryonic stem above the cotyledons.

Fertilization — the union of two reproductive cells to produce a new embryo.

Filament — the stalk of the stamen that supports the anther.

Flower — the reproductive structure of fruit-producing plants.

Funiculus — the stalk of the ovule.

Gamete — a haploid reproductive cell.

Gametophyte — in flowering plants, the haploid cells contained in the ovules and pollen grains.

Genome — the totality of the genetic information contained in an organism.

Genotype — the genetic makeup of an organism; the sum total of all the genes present in an individual.

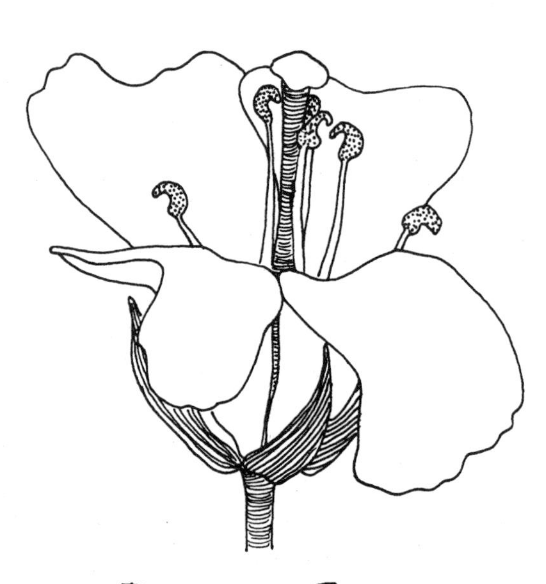

Brassica FLOWER

Glyoxysome — an organelle containing enzymes necessary for the conversion of fats into carbohydrates; glyoxysomes play an important role during the germination of seeds.

Gravitropism — the response of a shoot or root to the pull of the Earth's gravity.

Hilum — the scar where the developing seed was attached through the funiculus to the maternal tissue of the ovary.

Hypocotyl — part of an embryo or seedling between the cotyledons and the radicle or roots.

Imbibition — the absorption of water and subsequent swelling of the seed.

A longitudinal section of a flower

Incompatible — the condition of pollen and stigma that inhibits pollen germination.

Integuments — the outer layer of tissue around the ovule that becomes the seed coat.

Lateral meristems — meristems that give rise to secondary (thickening) tissue of the plant.

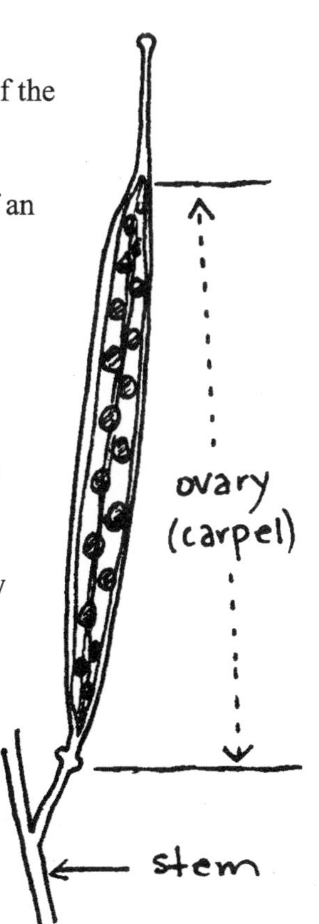

Longitudinal section — representation of a cut made parallel to the main axis of an object.

Male gamete — sperm (pollen).

Megagametophyte — the female gametophyte.

Meiosis — two successive nuclear divisions in which the chromosome number is reduced from diploid (2n) to haploid (n) and the segregation of genes occurs.

Meristem — the undifferentiated, perpetually young plant tissue from which new cells arise.

Micropyle — the hole in the ovule integuments through which the pollen tube passes on its way to fertilization.

Monocot (monocotyledon) — a plant whose embryo has one cotyledon.

Mutualism — symbiotic relationship beneficial to both organisms.

Nectary — the glands in the flowers of some fruit-producing plants located at the base of the pistil that produce a sugary fluid and attract insects and other animals.

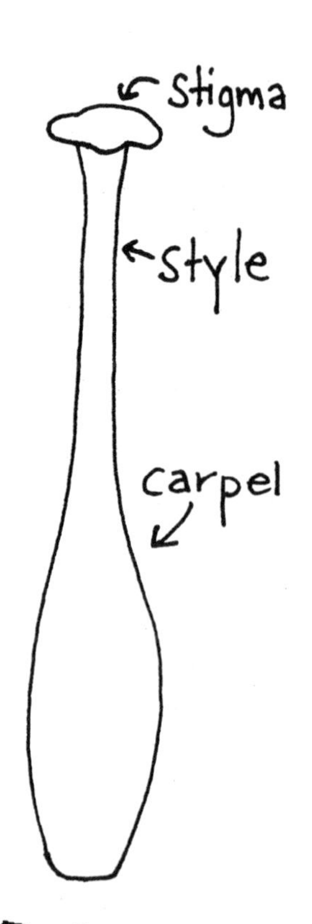

Oleosome (lipid body) — an organelle in which fat is stored in a plant cell.

Ovary — the enlarged part of the pistil containing the ovules. Also referred to as the carpel.

Ovule — the female reproductive cell in a seed plant.

Parasitism — a symbiotic relationship in which a symbiont (parasite) benefits at the expense of a host.

Petas — a flower part that is usually colored.

Phenotype — the physical appearance of an organism.

Pistil — the central structure of flowers, consisting of the ovary (carpel), style, and stigma.

Pollen — spores bearing male gametes; sperm.

Pollen tube — the tube formed after pollination that carries the male gametes into the ovule.

Primary growth — plant growth resulting in the extension of root or shoot.

Proboscis — any tubular structure for sucking, food gathering, or sensing.

Radicle — the embryonic root that first appears after germination.

Setae — feather-like insect hairs.

Scarification — the process of cutting or softening a seed coat to hasten germination.

Secondary growth — plant growth resulting in the thickening of plant tissue.

Self-incompatibility — the condition of pollen and stigma that inhibits the germination of pollen from the same plant.

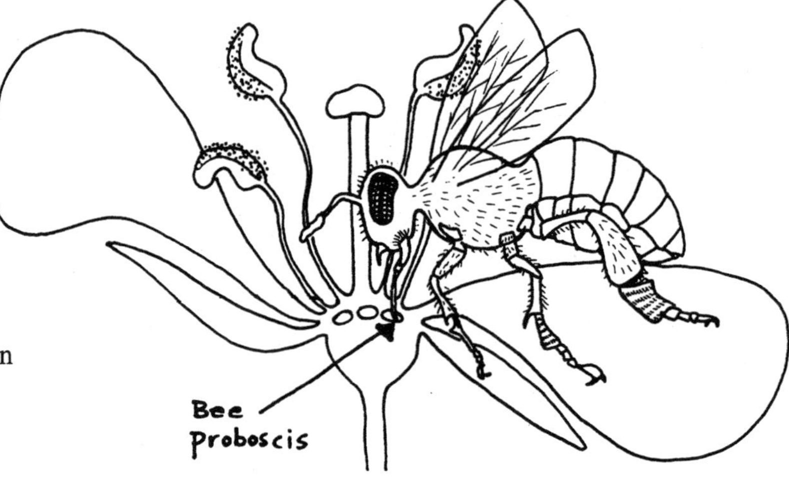

Senescence — the biological process of aging.

Sepal — one of the outermost flower structures. Sepals usually enclose the other flower parts in the bud.

Septum — a thin, paper-like partition separating the ovules in each carpel.

Shoot apex — the tip of the shoot (new growth above ground, consisting of stem and leaves).

Stamen — the flower structure that produces pollen; consists of the anther and filament.

Stellate — starlike, as in tufted clusters of plant hairs.

Stigma — the receptive surface of the pistil on which pollen adheres and germinates.

Stratification — the process of placing seed under a cold treatment, which will allow for subsequent germination.

Style — the slender column of tissue that arises from the top of the ovary through which the pollen tubes grow.

Suspensor — the strand of cells attaching the embryo to the ovule.

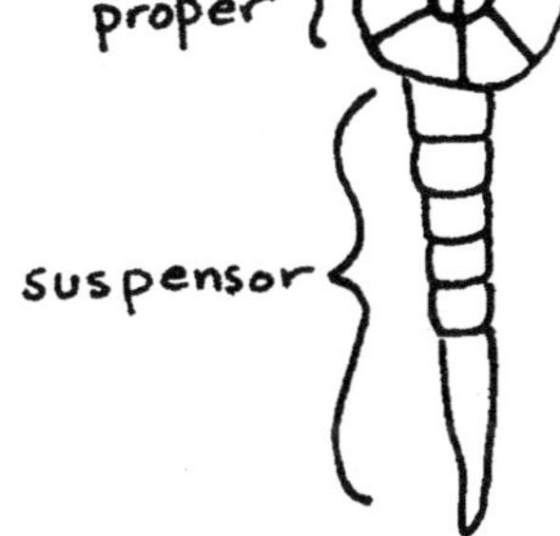

Symbiosis — an ecological relationship between organisms of two different species that live together in direct contact. Includes both parasitism and mutualism.

Testa — the hard outer coat of a seed that protects the embryo.

Variation — the differences that occur within the offspring of a particular species.

Zygote — the fertilized cell resulting from the union of the male and female gametes; the first cell of the new generation that develops into an embryo.

INDEX

A

Abscission, defined, 91
Anatomy, 2
  of flowering, 63
Anthers
  flower age and, 96
  open, 64
Anthesis, 65
Anthocyanin, 12
Apetalous seed stocks, 126
Apical meristem, 44, 95
Apidae, 79
Apis mellifera, 80
Arabidopsis, 43
Axillary meristems, 44

B

Bees, 78–80
  sketch sheet, 88
Beesticks, 80–81
  sketch sheet, 88
Biological diversity, 2
Brassica, flower parts, 63
Brassica rapa, defined, 1
Bud, young, 44

C

Canna compacta, 21
Canna seeds, 21
Capillary mat, 116
Carpels, 17, 62
Cell lineage, 44
Cell walls, 22
Cereals, 30, 98
Chlorophyll, 12, 22
Chloroplasts, 22
Class intervals, 128
Coleoptile, 22
Colors of light, 24
Compatibility, tissue, 79
Cotyledon, 12–13
  cells, 22
Cross-pollination, 15
Cross section, of flower, 67

D

Data analysis, 3, 37
Days after pollination (dap), 82
Days after sowing (das), 48
Dehisce, defined, 64
Development, 5
Dicotyledons (dicots), 22, 30
Dissection
  flower, 66
  seed, 28–30
Distribution curve, 130
Dormancy, seed, 26
Double fertilization, 92
Drawing to scale, 135–136

E

Embryo development, 94–95
Embryogenesis, 91, 96–98
Emergence, of seedling, 44
Endangered species, 21
Endosperm, 30, 98
Enzymes, 22
Epicotyl, 22
Evolution, 2
Experimentation, 3

F

Fast Plants
  as teaching tool, 1–2
  genome size, 43
  growing and maintaining, 113–126
  growing systems, 114
  growth curve, 18
  harvesting of, 122–123
  lighting systems, 114
  material sources, 125–126
  planting of, 119–120
  senescence in, 99
  time requirements, 114
  troubleshooting, 124
Fertilization to seeds, 5, 91–107
  background, 91–95
  embryo development, 94–95
  embryogenesis, 96–98
  pistil length data sheet, 104, 106–107
  seed dissection sketch sheet, 105
  stigma position data sheet, 104
  teacher page, 102–103
Film can hand lens, 133
Film can wickpot construction, 116
First flowers, data sheet for, 72
Floral diagram, 67
Floral parts, orientation of, 76
Flower
  age, 96
  buds, 13
  defined, 62
  opening of, 64
  spiral, 73–75
  unpollinated, 97

Flowering, 5, 61–76
  anatomy of, 63
  background, 61–62
  dissection, 66
  floral diagram, 67
  heating and, 65
  nectar and, 68
  orientation of parts, 67
  spiraling of, 66
  teacher page, 70–71
Fluorescent lighting, 114
Food reserves
  chart, 30
  in endosperm, 98
Form/function, 2
Frequency histogram, 128
Fuel molecules, 22

G

Genes, 2, 43
  senescence associated, 98
Genome, defined, 43
Geriatrics, plant, 99
Germination, 5, 21–41
  background, 21–22
  Canna compacta, 21
  colors of light and, 24
  cotyledon cells, 22
  data analysis, 37
  dissection of seed, 28–30
  imbibition, 26–28, 38
  observation data sheet, 36
  observation of, 23
  orientation of seed, 30–32
  sketch sheet, 37
  teacher page, 34–35
Germination chamber, 31
  sketch sheet, 40
Globular embryo, 94
Glucose, testing, 68
Glycoproteins, 79
Glyoxysome, 22
Gravitropism, 50
Grower's calendar, 19
Growing systems, 114
Growth curve, 18
Growth and development, 43–59
  background, 43–44
  genomes, 43
  hypocotyl hypothesis, 50
  meristems, 44
  teacher page, 54–55
  variation in, 43

H

Hair, on leaf margin, 48
Hairs, stellate, 62
Heart-shape embryo, 95
Heat, and flowering, 65
Hilum, 29
Humidity, 64
Hypocotyl, 12–13
Hypocotyl hypothesis, 50

I

Imbibition, 26–28
  data sheet, 38
Inquiry strand, defined, 3
Integuments, 93, 95
Internode, 14

L

Lateral meristems, 44
Leaf, young, 44
Leaf hairs, 48
Legumes, 30
Life cycle, stages in, 2, 12–17
  first 72 hours, 12
  days 4–9, 13

days 18–22, 16
days 23–26, 17
days 38–45, 17
day 45, 17
Lighting systems, 114–115
light house construction, 115
Lipids, 22
Longitudinal section, of flower, 67

M

Magnification, calculating, 135–136
Male sterile seed stocks, 126
Master activity calendar, 8–11
Mean, defined, 131
Meristems, 44
Micropyle, 28, 29
Model organism, Fast Plants as, 6
Monocots, 30
Mutualistic symbiosis, 77

N

Nectar, and flowering, 68
Nectaries, 62
Nodes, 14
Nucleic acid base pairs, 43
Nutrient solution preparation, 117

O

Observation, 3
Observation data sheet, 36
Oleosomes, 22
Organelles, 22
Orientation
of flower parts, 67
of seed, 30–32
Ovaries, 17
Ovules, 17, 62
per pod, 97

P

Parasitic symbiosis, 77
Peroxisomes, 22
Petals, 16, 62
Phenotype, 43
Photosynthesis, 22
Photosynthetic organs, 22
Physiology, 2
Pinto bean, 28–29
Pistil, 62
Pistil length data sheet, 104, 106–107
Planting medium preparation, 118
Pod, 16–17, 93
characteristics, 109
Pollen, 62
germination, 83–84
shedding of, 65
vectors, 78
viability of, 15
Pollination, 5, 77–89
background, 77–80
bees and beesticks sketch sheet, 88
beesticks, 80–82
by insects, 77–80
honey bees, 80
pollen germination sketch sheet, 89
teacher page, 86–87
Primary growth, 44
Proteins, 22
Pruning, 15

Q

Quantitative biology, 2

R

Radicle, 12–13
Range, defined, 131
Rattles, seed, 21

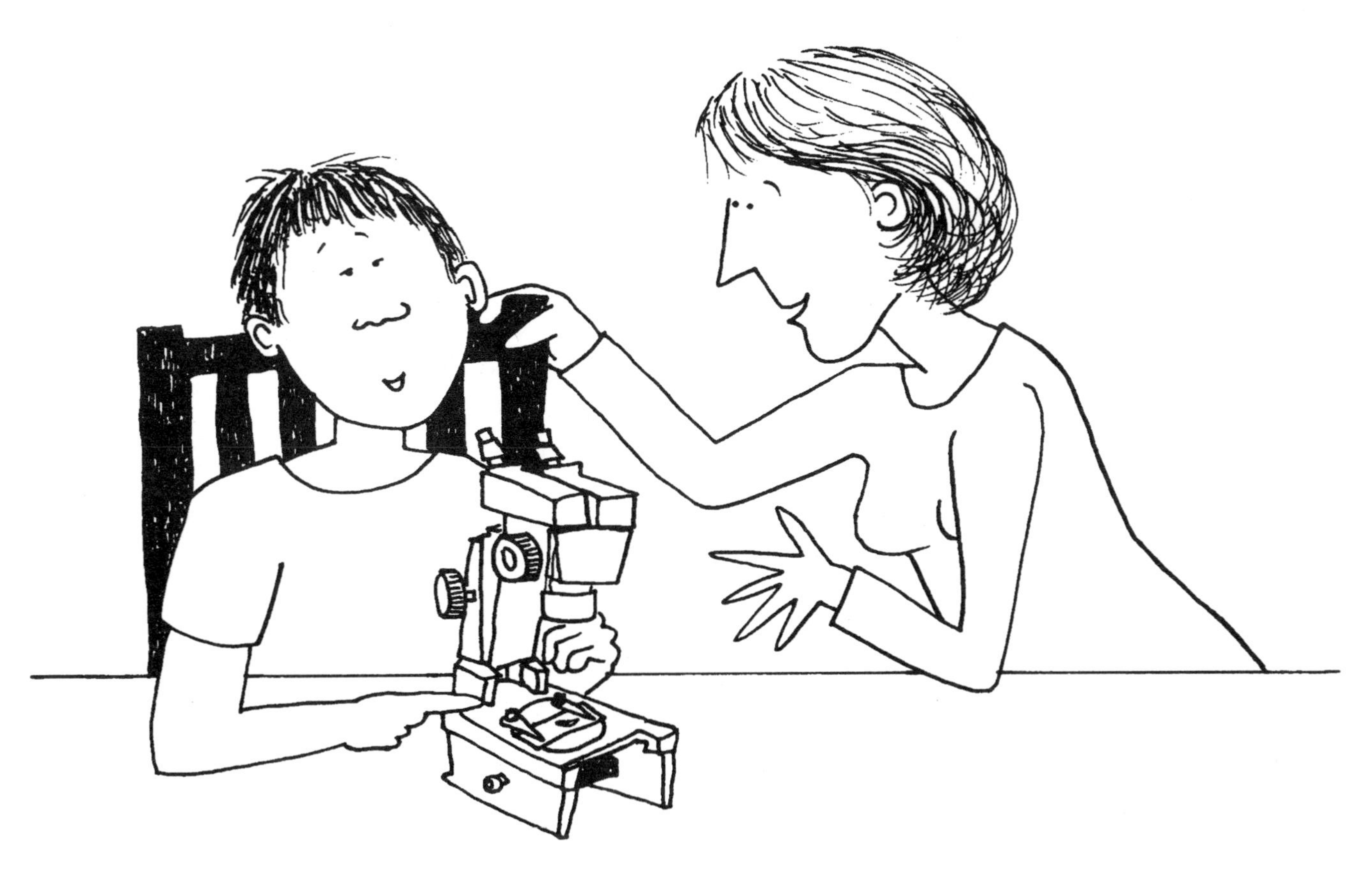

Representative sample, 128
Root apical meristem, 95
Ruler disk master, 41

S

Scale Strips, 135–137
Scarification, 26
Science standards, 151–152
Secondary growth, 44
Seed, 12–13
  dormancy, 26
  longevity, 21
  orientation, 30–31
  taste, 29
Seed dissection, 28–30
  sketch sheet, 39, 105
Self-incompatibility, of pollen, 79
Senescence associated genes, 98
Sepals, 62
Septum, 97
Shoot apical meristem, 95
Silique, 17
Solution concentrations, 24
Stamens, 62
Standard deviation, defined, 131
Starches, 22
Statistics, 49
Stellate hairs, 62
Stem elongation, 14
Stigma, 62
  flower age and, 96
  position, 104
Stratification, defined, 26
Suspensor, 95
Symbiosis, 77

T

Taste, of brassicas, 27
Torpedo embryo, 95
Trillium, 43

Triploid endosperm, 92–93
True leaves, 13

U

University of Wisconsin–Madison, 1–2

V

Variation
  genetic vs. environmental, 110–111
  tracking, 45–50, 56–59, 109
Variation, data on, 127–132
  amount of, 128
  interpretation of, 130–131
  organization of, 128–129
  type of, 128
Vectors, pollen, 78

W

Walking stick embryo, 95
Water bottle construction, 118
Wisconsin Fast Plants Program, 1–2

Z

Zygote, 94